How To Make A Living From Eggs and Poultry

by Herbert W. Brown

with an introduction by Jackson Chambers

This work contains material that was originally published in 1916.

This publication is within the Public Domain.

This edition is reprinted for educational purposes
and in accordance with all applicable Federal Laws.

Introduction Copyright 2017 by Jackson Chambers

Self Reliance Books

Get more historic titles on animal and stock breeding, gardening and old fashioned skills by visiting us at:

http://selfreliancebooks.blogspot.com/

Introduction

I am pleased to present yet another title on Poultry.

The work is in the Public Domain and is re-printed here in accordance with Federal Laws.

As with all reprinted books of this age that are intended to perfectly reproduce the original edition, considerable pains and effort had to be undertaken to correct fading and sometimes outright damage to existing proofs of this title. At times, this task is quite monumental, requiring an almost total "rebuilding" of some pages from digital proofs of multiple copies. Despite this, imperfections still sometimes exist in the final proof and may detract from the visual appearance of the text.

I hope you enjoy reading this book as much as I enjoyed making it available to readers again.

Jackson Chambers

Mountain View Poultry Farm

PREFACE

This little book has been written for the encouragement, as well as the warning of those persons, who starting in a small way, desire to make a living from chickens. I have "no axe to grind": no blooded stock or hatching eggs for sale, nor day-old chicks to dispose of. My business is the production and marketing of table eggs and poultry. Consequently I have no reason either to exaggerate the pleasant features or to withhold any unpleasant truths. The contents of this book are plain, unvarnished facts as I have met them. They are given for what they are worth to those desirous of building up a little plant of 1,000 or so hens.

H. W. BROWN,
Mountain View Poultry Farm.

Saugerties, N. Y.
1916.

TABLE OF CONTENTS

Chapter		Page
I	The Poultry Business	1
II	A Little Personal History	12
III	Qualifications Necessary to Success	21
IV	Selection of the Location for a Poultry Plant	28
V	Starting and Capital	35
VI	Poultry Houses Recommended	43
VII	Breeding and Hatching	71
VIII	Incubating	76
IX	Pointers on Hen Hatching	85
X	Feeding Chickens of Various Ages	89
XI	Fattening Broilers and Old Hens	101
XII	Water	105
XIII	Marketing the Eggs	108
XIV	Home Made Cases	119
XV	Storing Surplus Eggs	124
XVI	Tests for Fresh and Stale Eggs	129
XVII	Marketing of Broilers	131
XVIII	Selling Off Old Hens	135
XIX	Wire Runs and Delivery Crates	138
XX	Cost of Egg Production	140
XXI	Characteristics of Breeds	142
XXII	Breeds and Weights of Fowls	144
XXIII	Growing Own Feed	145
XXIV	Common Diseases of Chickens	147
XXV	Routine Work in the Spring	152
XXVI	General Remarks	155
XXVII	Don'ts Learned from Experience	165
XXVIII	Conclusion	171

LIST OF ILLUSTRATIONS

	Page
Mt. View Poultry Farm	Frontispiece
The Part of the Poultry Business so Alluring to the Beginner	4
Partial View of Farm from the Top of Laying House	14
One of the Routine Duties of a Conscientious Poultryman	24
Front View of Part of 128 Foot Laying House	44
Laying House, Interior View	47
Detail of Front of 47 Foot Section of Laying House	50
Floor Plan of Same 47 Foot Section	51
Detail of Nests	52
Detail of Ends of Laying House	53
Home Made Fireless Brooder	56
Same, Showing Corners Rounded	56
Simple Fireless Brooder	57
Colony Brooder House Showing Hover	58
Details 6 x 6 Colony Brooder House	59
Commercial Hovers	60
Water Pan, Feed Trough and Feed Hopper	61
Details of Colony Roosting House	62
Details of Roosting and Brooding Coop	63
Hen Brooder Coop	64
Details of Hen Brooder Coop	65
Small Colony House	66
Brooder House, Half Underground	68
Details of Brooder House	69
Floor Plan, Brooder House and Incubator Cellar	69
Hover Construction in Brooder House	72
Egg Turning Case	74
Little Chicks Feeding at Covered Trough	90

ILLUSTRATIONS

Covered Water Trough	91
Home Made Single Feed Hopper	92
Colony House with Wire Run	93
Home Made Double Feed Hopper	96
Details of Double Feed Hopper	97
My Method of Supplying Water to Chickens	106
Details of Home Made Egg Cases	120
Dimensions of Egg Crates, Cost and Cost of Shipping	121
My Method of Packing Eggs	122
Wire Run	139
Combination Delivery Crate and Wagon Seat	140
Raising a Colony Coop to Avoid Rats	158
Hen Catching Crook in Use	161
Feed Trough for Growing Stock	166
Chickens Feeding at Covered and Uncovered Troughs	168

INTRODUCTION

A LIVING FROM EGGS AND POULTRY

The book is not intended to be a scientific exposition of the chicken business further than a diagnosis of the business methods applying to the business if going into the details of those methods, might be called scientific.

It is possible to make a success of the chicken business without any knowledge of the chemical analysis of the egg, the chicken or the chicken feed.

It is conceded of course that the more knowledge one has of every possible detail entering into his business the more likelihood is there of that one's success; but we must creep before we walk.

No one can expect to be thoroughly conversant with the practical as well as the scientific elements pertaining to a business at the time he first enters that business, and for this reason business of any kind becomes more interesting.

The happiness of life is derived from ever reaching out from one pleasure or attainment to a new one.

INTRODUCTION

Life would indeed be monotonous if we had all our desires and ambitions gratified at once: so with business—unless one takes real pleasure in his work the success of that business is doubtful but as each day unfolds something new, some added knowledge to be gained, renewed interest urges one on to the attainment of higher standards and more successful work.

I read one work on Poultry which filled over 1000 pages of very interesting and instructive matter but from which the beginner would have been able to get but little of practical value.

It was like algebra to the student who has finished arithmetic and moves forward into a higher plane—but the arithmetic, the practical value had to come first.

Likewise in the chicken business, the novice must have a general understanding of the practical part first, and then add all possible knowledge that will make the business more interesting.

I have written this book in the hope of aiding beginners in the chicken business and mean by "beginners" young men from the city like myself, who, when they come to the country are as ignorant of rural life and

INTRODUCTION

all matters pertaining to country life as I was, who did not know a Leghorn from a Wyandotte.

Why I had a friend who was a college graduate say to me not long ago that he intended buying a place on which to keep chickens and to avoid having so many young cockerels in the summer he would order his day old chicks to be sent 75 per cent. pullets and 25 per cent. cockerels, and I really wonder how many city men whose thoughts might wander to a poultry farm to be his some day, would know if he was right in his theory?

Keeping this fact always in view, that a novice is not supposed to be familiar with any of the details of his new business, I have explained more fully the minute details usually omitted from other books, but which are really the most important phases of poultry keeping and have endeavored to make my story direct and to the point without wasting unnecessary words or including a vast amount of matter that a novice could not master at first, but leaving that for his future study.

A Living from Eggs and Poultry

CHAPTER I

THE POULTRY BUSINESS

So much is now being said of the poultry business; so many articles are being contributed to the various papers giving experiences of some persons who have made good profits from small backyard flocks of hens; so many advertisements are daily appearing in the magazines using the successes of the business, to boom the sale of farms, poultry appliances, day old chicks, etc.; and so many alluring accounts of wonderful, immense poultry farms making thousands of dollars a year profit, that a large proportion of our population must believe the easy road to prosperity is in keeping poultry.

I am by no means a pessimist, but I believe as do all these interested writers and advertisers, that there is money in chickens: not barrels of it or a gold mine, but a living, good or bad according to the man himself who develops the business. But I also believe it is time that some one should give to the public the facts as he finds them in establishing a business with 1000 hens—a plant such as the average man would think to be of fair size.

I love the business, but believe I can help the poultry industry more by giving the facts as I have seen them, as they have been driven home to me, as, after having read and planned and worked, I have either failed or succeeded, than by exaggerating these facts or referring to the large successful plants of years development and large investment. The latter methods of description too often infer that any one can at once do the same as the owners of these plants did only after years of experience.

What has been done can be done again. A man can set his ambition as high as he will, the higher the better provided he keeps his feet on earth, realizing that there are many steps between his beginning and the realization of his hopes and ambitions. It is those first steps, (which perhaps for more than one reason the average advertising chicken book does not go into too deeply) that the beginner should acquaint himself with and which this little book will bring to his notice.

Most of the help and inspiration derived by us from reading the history of some great man's life comes not from the fact alone that his was a successful life, but from the fact that his success was attained by overcoming obstacles such as poverty, the lack of an early education or some physical affliction.

For this reason any poultry book addressed to beginners, written to induce novices to start in the business, a book which points only to successes, but neglects to warn of possible failure, and omits the early steps and probable mistakes, the hard work and the disappointments which must all be met and overcome before the success is attained, does not do justice to the reader who may be debating his chances of success in the business.

TOOK UP BUSINESS IN QUEST OF HEALTH

I came to the country as many others have come and will come, looking for health after having lost the greatest of God's blessings in the strenuous city business life, that men are daily feeling more and more burdensome. Being located near a summer resort I came in contact with a great many men, a majority of whom seem to have the "back to the farm" idea with a preference for poultry. From conversation and reading, I know a great many men have in their minds ownership of a poultry plant sometime in the future.

The part of the Poultry Business so alluring to the beginner.

Most people believe the poultry business an easy and sure way to prosperity. It is to the same extent that any other business is when properly conducted by a man willing to work. But easy—never; for there are as many details and annoyances connected with raising chickens in large numbers as are found in most lines of business; indeed perhaps more.

"Is there money in chickens and can I make a living by keeping a flock of hens?" Every poultryman has been asked this question so many times that some have answered in books usually in the affirmative and with glowing accounts of big profits—one as high as $6.41 a year a hen!

The profits a man can make a hen are governed by a great many things such as the particular market he may have, whether he can sell his surplus hens and roosters at a high price, or his eggs for hatching or day old chicks at extravagant figures.

I have selected the utility branch of the business, the supplying of table eggs and poultry, and will endeavor to show how, during three years I have averaged from $1.50 to $2 a hen a year and how it will be much safer for the beginner to take these figures, which he will find many others will approve rather than the extravagant estimates of so-called "experts," under conditions which the beginner cannot hope for, at least for several years.

COMPARATIVE INCOMES

First of all, what is your income that you purpose to give up to enter the chicken business, and what are the profits you have in mind that you expect to get from chickens. A thousand dollars a year? Then you must plan to keep a flock of 500 to 750 fowls. Two thousand dollars a year? Then figure that your plant will have to number 1000 to 1500 hens. On this basis you should be fairly safe and could then study the special methods used by these experts who make the vaunted large profits, building your business and dividends up to these higher estimates.

You must understand that the poultry business, at least the table egg branch, is a business made up by the multiplication of small units; that you are not in the line of high finance but engaged in producing and selling very small things. Eggs are sold at 2 to 5 cents each provided you are so fortunate as to sell any at 5 cents. Even if you do you will find at that time they are worth the price, as it is simply the lack of supply that raises the price.

The ordinary, good hen will not average much better than one egg in three days the year round when kept in large flocks. Averaging these eggs at 2½ to 3 cents each, you will find your hens manufacture about 1 cent a day. Isn't this a small business? It is only in numbers, therefore, that you can expect to obtain a fair living. The young man expecting to reap a harvest from chickens must remember he will require a flock of 500 at least, besides the equipment to keep it, before he can even expect a fair income. If he hopes to add to his plant it will take some of the income to do that. All such details will be gone into at length later in this book.

When eggs are quoted at 4 or 5 cents each it is only natural that the uninitiated should think there is lots of money in chickens, or that they are being mulcted for the poultryman's benefit; but if they could look in on the poultrymen at this time they would see the majority gathering about one-tenth of the eggs gathered when the price is 2½ to 3 cents each. At this time it is discouraging to know you are feeding a dollar a day and getting only 10 cents in return. Then it is that Mr. Poultryman needs a great gift of patience and of hope for better things to come, and it requires determination of a higher order to work as faithfully as when the hens are making some returns.

Another necessary thing at this time is some cash saved from spring and summer or a little nest egg saved from some other source. Some of the encouraging features of the business, however, are as follows: The flock of hens can always be disposed of at cost or at a very slight sacrifice unless it is of a fancy breed, in which case perhaps, this statement might not apply. In the starting of how many other lines of business can the investor be assured that, if at the end of the year he wants to sell at cost, he can do so even without getting any dividends from his stock as he would get from his hens? Every man launching a new business must take into consideration the fact that he or his heirs might wish to sell. It is a good business indeed that at forced sale would return anything like the amount of money invested.

Another of the safe, encouraging features is that of the ready market for all that is produced. Compare this advantage with those of other lines. What a snap a manufacturer would feel he had if he knew he was certain that no matter how much he produced he could not glut the market and that without expensive salesmen he could box and ship his product at a fair price.

POULTRY A BILLION DOLLAR BUSINESS

The New York Tribune of October 13, 1913, stated that 1,700,480,880 eggs went into the New York market in 1912. Is it likely that the few you and others will produce, will affect this market or the price at which they were sold? This is a billion dollar a year industry you contemplate entering. For some years at least, you need not worry about market and price, at any rate, not until the movement of young men is toward the country rather than the city. This rapid growth of the cities out of proportion to the development of the country is the main factor causing high prices.

Each year by the thousands young men are turned out by the schools and colleges to face a business career for which they are supposed to have been preparing, and each year these thousands find it more and more difficult to obtain lucrative positions, because of that never failing law of supply and demand. This is the very law that adds to the chances of success of the poultryman, because, as the cities fill up with their vast populations which have to be fed, so the opportunities of the poultry business increase.

The wages or the salaries of the men we speak of—not mechanics, who partly through their labor unions and partly through the young men selecting professions or clerical work, have had increased wages—have not advanced in proportion to the advance in living expenses. But how about the cost of food stuffs—have they advanced? You know to your sorrow that milk, butter, eggs, meat and all other farm products have kept climbing up in price. How can it be otherwise when 99 out of 100 young men choose the life of the consumer rather than the life of the producer. Prices will remain high in spite of "trust busting" and lower tariff which, however, may have some slight effect toward a reduction. Until more young men realize the opportunities awaiting them in the food producing lines the trend of prices will be upward rather than downward.

The young men, who, seeing the opportunities, studying and gaining all the knowledge they can from others and then entering these fields of production, will be at least as well off at the end of five or ten years as their fellows who selected the occupations offered by the city.

Last but not least of the favorable features of the poultry business is health. It is a great question whether it was God's plan to have the cities of vast population, with palaces and hovels, great hotels and apartment houses, gigantic office buildings, factories and sweat shops, with the awful crowds on all the transit lines—while great stretches of open country are lying waste. "God made the country and man the cities" is a well known proverb. In following God's plan the young man will certainly make no mistake.

But, another word of advice: Make sure your family will be satisfied and contented with the life you are contemplating. A great many women have an antipathy to country life after having been accustomed to the conveniences, pleasures and luxuries of the city. This is largely a matter of disposition, depending upon what one needs in the way of pleasure and what is his or her idea of the requisities of a full and perfect life. Fortunate is the man, who, inclining toward the country and its work, has a family whose tastes are the same and who would take pleasure and enjoyment from and in his work, thereby encouraging him in his ambition.

CHAPTER II

A LITTLE PERSONAL HISTORY

As my decision to go to the country was forced upon me some personal history will follow, including the steps in the building up of my little business. From it the reader can see why I am able to give advice to those who would take up poultry keeping in a small way. It is to and for this class only that this book is written. The experienced poultryman has been all through practically the same experience but in his ads and writings he says little of anything but the final success. A great majority, however, of the novices desiring to commence business would have to begin in a small way, working up just as I have done. With an honest, unbiased history before them of failures and mistakes, as well as successes, and the things I have learned by profiting by those errors, these beginners will be able to avoid some of the pitfalls, thereby reaching success by a shorter route.

A LITTLE PERSONAL HISTORY

I was not dependent upon chickens for an income on which to live; but, having some ambition left and looking toward the future, and wishing to stay in the country, chose the keeping of chickens because it appealed to me. Not knowing from year to year whether my residence would be permanent I did not try to increase my flock rapidly until the last of the three years, consequently the reader must not think that a larger flock than 800 can not be developed in the three years.

The beginning of the plant consisted of six Leghorn fowls purchased in February, 1910, and housed in a home-made coop 3x6 feet. These fowls began to lay shortly after and the chicken fever began apace so that 40 more were bought, giving me not exactly a mongrel stock but still a flock of white and brown Leghorn, black Spanish and Wyandott fowls of no particular strain.

COSTLY MISTAKES THAT TEACH LESSONS

At this point note the first mistake and profit by it: the starting with a more or less scrub bunch of hens. This is not good policy. Rather decide upon the breed which suits your fancy best, or will best produce the results for the market you purpose to

Partial view of Farm from the top of Laying House.

A LITTLE PERSONAL HISTORY 15

supply. Then get a fair strain of this one breed, study its peculiarities, find out its good and its bad points and try to improve it. (See starting business, page 31.)

This flock of 46 laid well, so that some eggs were sold but not at good prices, for the flock was so small there was really no use looking for a good market. An incubator was bought, also a few day-old chicks, and the summer was ended with 45 hens and 135 pullets. In the fall a 40x15 foot laying house was built and during the winter a 20x15 foot brooder house was attached.

Again note the very serious mistake of building an incubator or a brooder house in conjunction with any other house, which my experience will bring forcibly to your mind. When February came around 65 nice little fellows were hatched and were six weeks old March 29 when 160 more hatched and were put in the hovers. At seven o'clock the cry of "fire" was heard, and by eight o'clock the entire plant was in ashes excepting 65 hens and pullets which were saved.

Imagine, if you can, my feelings. In poor health, on the eve of what appeared to be a successful early season, to lose 225 little chickens, 110 hens, two incubators, etc., and a poultry house just completed; a loss totaling $500 less $150 for eggs and fowls previously sold. All caused by an error in judgment!

It took me two days to get over the shock and the disappointment. But on April 1 a new account book was started with the 65 survivors, some of which looked as though they had been through the war. A 244-egg incubator was borrowed and work begun again with the result that at the end of the summer I had a goodly flock of chickens.

Having acquired a two-acre plot in a different location, a 20x15 foot house was built for the 65 hens in July, and the little ones moved to new quarters where they did very well, necessitating the building of an additional 60 to the 20 foot house occupied by the hens. About 225 pullets were moved into winter quarters in October, being kept separate from the hens so the year's record of these 65 hens could be kept straight. This record at the end of their year showed just $1.50 profit on each hen.

The flock was now of fair size. Realizing that soon quite some eggs would be coming; in January a few dozen were sent to a friend in my home town. These being quickly disposed of, led to a steadily increasing demand, which kept pace with the increased supply in the spring, when 60 dozen a week were being shipped. Meanwhile a 42x15 brooder and incubator house had been built, but this time separate from all other buildings. In February two incubators were set going. Even with very poor hatching results, the flock grew some. After disposing of the old hens and the scrubs I finished the season with 400.

DECISION MADE TO ENLARGE BUSINESS

Having decided to remain in the country, the previous work was carefully analyzed, mistakes noted and profits examined with the result that I decided there is some money in keeping chickens, but to amount to much, the business would have to be increased. Feeling that the previous two years' experience was sufficient to warrant my keeping more hens, I set out in the fall to enlarge the flock to a respectable size the next year.

Not having had very fertile eggs for hatching the previous winter, incubators were not set until March 1. It seems to have been just as well for the hatches were larger in percentage, the chicks seemed to be stronger, losses were less and altogether things moved along very smoothly resulting in the hatching and rearing of about 900 chicks, of which about 450 were sold as broilers and the remaining 450 pullets added to the flock, which then totalled 800.

A 40-foot addition was built on the laying house making it 120 feet long. The extra eggs from the flock of 400 were sold to the home customers and to one hotel that took 30 dozen a week. In eight months this flock laid 33,000 eggs, all of which were sold at fair prices, only a few being sold at 28 cents a dozen, the balance from 30 to 45 cents. I feel now, that the hardest part of the task of building up a little business has been accomplished and that it will be comparatively easy to bring the flock up to 1,200 or 1,500 during the fourth year. With a continually increasing market, I do not hesitate to aim for that number.

The following pages will be devoted to enlarging on particular features of the business met in the starting and development of my plant. All of these points will have to be met and solved by every man beginning the business. Profits, cost of houses, selling of the eggs, capital needed, mistakes made—just the things that a man with the "chicken fever" wants to know, will be gone into from the standpoint of a beginner, with the hope that the figures given and the pages written will be of more help and encouragement than if written by some big poultryman with 5,000 or 10,000 hens, which sized flock the majority of men will never have.

I have stated that I have made $1.50 to $2.00 a hen strictly in the utility line, i.e., the selling of table eggs with broilers as a side issue. While much higher figures are given by some, I will show where my figures come from.

INCOME FROM THE FLOCK

April 1, 1911 to April 1, 1912—60 to 65 hens netted $95.

1912—275 to 300 hens netted (including increase) $336.

1913—400 hens netted (including increase of 400 spring pullets, worth $1.00 each) $800.

The plant has cost for fencing, incubators, brooder house, laying house, colony houses, etc., $1,075; but to offset this amount there are 800 hens and pullets and $700 cash profits. However, no salary or interest has been deducted from the above, as it would be difficult to figure interest on the investment on a continually developing plant; nor salary either, until the plant has grown to a fair size, when I can credit profits to the account that suits me.

CHAPTER III

QUALIFICATIONS NECESSARY TO SUCCESS

I will presume the reader desires to venture in the poultry business and in the branch of the business that has to do with the selling of table eggs and poultry. The qualifications necessary to make a success are practically the same as those needed in the establishment of any other project, but perhaps with some little differences.

1. **General Ability.** It seems to me that it requires more real ability to develop a business from nothing up to a one of good size than to conduct such a business after it has been built up. In the first place a small business, in its inception, will not stand anything like the over-head charges of a large business. System, is one of the important factors in conducting a business economically, but system is not so easy to apply when one head has to do all the planning, etc., and one pair of hands all the little varieties

of work, as it is when each particular part of the work can be apportioned to different hands, and the buying, selling and producing can be attended to by different heads. Supplies also can be bought more cheaply in the quantities a large business requires than they can in small lots. So do not think because your poultry farm may be a small one that you do not need a fair general idea of business; for you will be, in a way, a manufacturer as well as a retailer of eggs.

2. **Capital.** There are two ways of starting the poultry business as regards capital. One way would be to buy an established farm which is paying large enough dividends to support the purchaser. This would be practically the same proposition as the buying of any other paying business and would simply be a question of whether the purchaser has the capital and the ability to conduct the new business. The far larger class of beginners would have to start the other way—with a limited capital. They would expect and hope by their own endeavors, to build up the larger business. This would probably be the safer way, besides being the way leading to the greatest pleasure and satisfaction.

Anyone with money can buy a business, but not everyone can build up a business even if he may have money. One of the most satisfying elements of productive work is to see a flock of 200 hens the first year increase to 400 the second, and 800 the third, with more houses and more profits, and to know that each year's growth and added prosperity is the result of his own handiwork and study, which have added to his experience so he is ready for larger things.

But even commencing in a small way requires capital. Here is where many enthusiasts may make a mistake which may lead to discouragement or worse. I have stated that it is risky for a novice to count on more than $1.50 a hen. It would be equally hazardous to begin with a large flock the first year. Perhaps this might result in loss and discouragement. Here then is where capital is required to tide over the first year which any sensible man embarking in a new business, would expect to be more or less experimental.

Use the estimates which will be elaborated on later, and the yearly living expenses, to estimate how much balance to have at hand. You will certainly need it. Twenty dollars may seem small in a pay envelope; but quite a flock of hens is needed to return that much average weekly from eggs and broilers. Remember this and save yourself worry and trouble.

One of the routine duties of a conscientious poultryman.

3. **Willingness to Work and Stick.** Most city bred men work hard and more steadily than country men, but the work of the poultryman is of a different class from office work. The man looking toward the chicken business must realize that old clothes will be his regular attire and that farm work cannot be done in a neat business suit.

The work is not all in gathering eggs. There is plenty of dirty work, such as cleaning, whitewashing and disinfecting houses, digging post holes, cleaning chickens, etc.—work that some men would probably have to cultivate a taste for, but which must be done to keep the hens in condition to do their part. None of the work is heavy, but commonplace, the reward being the satisfaction of knowing that the more of this kind of unpleasant work done, the more chances of success.

With willingness to work also goes the faculty of taking failures and disappointments philosophically. The experienced as well the novices are sure to meet with misfortunes because they are engaged in dealing with animals that may contract disease or may meet with accidents—animals with wills of their own, and very perverse wills at that.

"For ways that are dark and tricks that are vain
The Heathen Chinee is peculiar,"

but the average hen has him beaten a mile. She will get through a 6 inch hole but in driving her back will pass by a 3 foot gate and not see it. You can supply her with a nice clean nest but she will look for every other conceivable place to lay her eggs, even on top of the curtain fronts of the houses, or she will set on a couple of china eggs until they are hot, but leave a nice setting of 15 real eggs (perhaps worth $5!) to hunt up those china ones, and if they have been taken away she may sit on the place that recalls them to her memory. The roosts may be exactly right, but if she prefers a tree or a window sill or the side of a nest box it will try your patience to make her see things your way.

"How use doth breed a habit in a man"— and in a hen. But, thank fortune there is one good feature in this; for once you get her into your way of thinking your troubles are over, for "chickens come home to roost."

4. **Market.** A highly important feature of a successful poultry business is a good market for produce. In the first place make your decision as to what kind of market you will cater to and bend your energies to that end. Business today is a matter of specialization. Few men can meet the severe competition of the times and be successful, unless their energies are set on one thing rather than divided among many. So a beginner would be safer in deciding at once that he will build up an

QUALIFICATIONS FOR SUCCESS

egg and broiler business, a fancy breeding business, or a day old chick business, rather than to combine two or more of these special lines.

The whole plan of developing the various branches is different in details as to growing, selling and advertising. As my experience is based solely upon the utility branch my remarks refer entirely to that line. Market will be gone into detail under advice as to starting poultry business.

5. **Adaptability.** To be gifted mechanically is a fortunate asset, as there are numerous little pieces of carpenter work to be done in the way of hoppers, nests, houses, egg boxes, etc. In fact, until a plant is well built up there are endless things, that if they can be done without employing a carpenter, will mean a considerable saving.

CHAPTER IV

SELECTION OF THE LOCATION FOR A POULTRY PLANT

The selection of the location in which to commence poultry raising is a matter of many phases, such as proximity to market to be supplied, value of the land, intended investment, or inclination as to open country or more urban life. The ideal location is usually described as land sloping gently towards the south, dry and yet with a stream of water flowing through the property, some shade and some sunshine, etc.

If one can find a location embracing all these advantages he is fortunate, but practically the only really essential feature is that the ground be dry. Wet ground might do for ducks but dry soil is necessary to keep chickens. Shade is a great advantage, but can be supplied artificially if there is not an abundance of it naturally. The buying or leasing of a place is a matter about which outside advice does not count for much, except that a man should feel sure he will be satisfied and will be able to make a permanent business before he buys.

But location as to market or more particularly as to the shipping point is very important. A retail merchant usually selects his place of business in the heart of the retail section regardless of the rent, but the manufacturer selects his more as to shipping facilities and the saving of rent. While the poultry business is really both manufacturing and retailing still the producing end is of most importance in the matter of location, for the eggs may be produced on land worth $25 an acre just as well as on other land worth several hundred. A crate of eggs or poultry leaving farms 20 miles and 100 miles from their destination in the afternoon would arrive together at the market the following morning; but the 75 miles might mean considerable difference in the expenses of the land on which the plant is located.

Again the greater distance from the city market might bring one in contact with a summer colony, in which case the product in the summer, when city people usually write, "Do not send any more eggs until September," would meet a ready market at good prices. My experience with the summer market is such that I would advise the beginner to look well into this feature before settling upon a location. Most novices would probably not invest the first season in sufficient heated brooder houses, etc., to be able to supply any early broiler trade, the majority of 2½ pound cockerels reaching that weight in July and August and even later.

The modern scientific way of hatching is to get early chickens in February and March, or April at the latest, and thus bring the pullets to maturity in September and October. This is really the only way that pays, but the beginner will probably keep on hatching later because his early hatches will not have been good. This will give him a lot of little cockerels late. If these can be marketed alive in July and August, near home in small lots just as the proper weight is reached, he will be fortunate. The same rule applies to eggs as well. Nearness to the railroad as the shipping and receiving point is important. If you are near a village you will find that very convenient also.

Considering these two points together see what you would save. Since 500 hens require about a ton and a half of feed a month, if your location is near enough to a village to have this feed delivered without charge you would save approximately $3.00 a month. Again, if you are located in a village you will not have to keep a horse to deliver shipments to the station, which shipments might total 100 or more a year and would cost considerable in time as well as money.

Then there are all the other supplies to be purchased—lumber, hardware, paint, etc.—which being needed continually may cause delay and annoyance if you are far from the village. All these points, pro and con, must be weighed and your decision made as your judgment sees right. My farm is seven miles from a village and I know whereof I speak. Even with these disadvantages I believe there is some money to be made with chickens, but the profits are affected somewhat by the aforesaid features.

WHAT MARKET TO CATER TO

Choice of location having been decided, you should decide upon the market you are to cater to and buy your stock accordingly. For instance, we will take table eggs with broilers as a side issue, always as the basis of our writing. You should have a flock of hens laying white eggs if you are to ship to

New York, because the quoted prices are higher on white eggs than on brown ones. This shows a preference of the purchaser for these eggs. The color of the shell does not affect the quality of the eggs. The food the hens eat does have an influence. If your prospective customer prefers white, it is folly to start with hens which lay brown eggs.

There are several breeds that lay white eggs: Leghorn, Black Minorca, Black Spanish, etc. Most egg farms producing white eggs use the White Leghorn, which is generally conceded to be, for egg production alone, the best breed. If brought through an early molt or if hatched and raised as pullets at the same time as other breeds this breed will lay as many winter eggs and in the spring, with much less inclination to hatch, will exceed the others. Hens of the heavy breeds prefer to set and brood chickens in the spring than to lay. If you have many of them and do not want them to set, it will keep you busy trying to break them of their motherly habits. This is annoying to you as well as to the hens which want to lay.

Again Leghorn cockerels can be brought to 2 or 2½ pound broiler size, at least as soon as the heavy breeds. At that weight they make as choice eating as any, as their bones are smaller and their frames lighter and therefore more meaty. Their activity also appeals to me as they are great workers. While it is sometimes provoking to see an unclipped hen sail over a 7-foot fence, still it is preferable to the slow action of the big ones. A good many people have said to me that they wanted a breed of fowl that when butchered to eat, would weigh more than the Leghorn, but if one is in the egg business the meat part is of secondary importance. A fowl is killed but once. At the end of two years one of a heavy breed would return in flesh two or three pounds more than a Leghorn—worth 35 to 50 cents—but if the Leghorn would not return more than 25 cents a year in eggs, my experience doesn't count for much.

VALUE OF WELL KNOWN STRAIN

Here is a good place to call attention to one feature of the business whereby some poultrymen increase their average profit a head; that is, by having a well known, well advertised breed or strain. They are thus enabled to sell those hens at the end of what they consider the most profitable laying term, for an advanced price over the meat price. You must understand that hens past the pullet age begin to stop laying in August because nature requires that they rest, shed their old feathers and put on their new winter plumage. This molting season is a trying period in a bird's life. The hens lay fewer and fewer eggs, finally none at all, if they are molting fast, until they have acquired their new feathers. This consumes two to three months.

At the beginning of the molting season the poultryman decides as to the hens he will dispose of and sells them either as meat, or breeders if he has a well known breed. If sold as meat Leghorn fowls would return 50 to 75 cents each; but if you have this other market, from $1.00 to $2.00. Unless you have spent considerable money for your first stock and have done some advertising you will find that it is not as easy a proposition as it would seem. If I were looking for 100 breeding hens I would buy first hand from the experienced, well known fancier. For this reason the beginner should be careful how he takes the statements of the successful breeder and compares or uses them as a basis of figuring his expected profits in the start; otherwise disappointment is likely to follow.

CHAPTER V.

STARTING AND CAPITAL

Presuming that you have decided upon a location and a breed, my experience as to the best time to start, would be in the very early spring, with a fair number of yearling hens or pullets. Hens in the spring are fairly easy for even the inexperienced to care for. Winter with the necessary confinement of the hens, with the consequent liability to disease, having past, it is to be presumed that whatever hens one could get at that time would be ordinarily healthy. As spring weather approaches they would be fairly safe from little mistakes one might make. Spring is nature's time of re-invigoration and life generally does better at that time than at any other season.

The second reason for starting in the early spring is that from then on for six months you would not only get the experience of taking care of the hens at the easiest period, but from the start you would begin to get considerable revenue in the way of eggs to pay not only for the keep of the hens but also for raising a considerable number of young chicks.

It seems to me that a man, dependent to some degree upon the new business for revenue, to start business with no hens to help his income, nothing but day old chicks from which he would get nothing for three months when he could begin selling a few broilers, then to wait three or four months before his pullets begin to lay even in small numbers, but meanwhile paying out money for feed, houses, etc.—that man would have to be a dyed-in-the-wool optimist to keep up his courage the first year. That first year is the hard year always in this business or until you have say 500 fowls. Then the outlook begins to brighten as you really see some chance of making your living from chickens. If you hold on up to this point you will probably stick.

Provided you think these arguments correct, buy say 200 fowls for which you have already made provision as to housing. If you expect to breed from them, choose a fair strain, but almost anything if you do not.

My idea as to building up a flock rapidly is not to depend the first season upon incubating your own chicks, as incubators in the hands of the novice sometimes do not turn out the results hoped for because of inexperience, infertile eggs, accidents, etc Rather buy one incubator with which to experiment, but make your estimates and calculations entirely outside of the incubator, buying whatever day old chicks you can afford. Let these day old chicks be your foundation stock, good, hardy chicks of a fair strain of the one breed you want to develop. Don't buy several breeds; make your choice and stick to the one kind.

It takes the average beginner with the average flock of pullets, six to seven months to get the majority laying. There is more money in having them all matured and of laying age by December 1 than to have part of them stringing along to February; so it would be advisable to buy all stock in March and April. This would assure you of the early matured pullets. If you count on incubating your chicks yourself you will find that the first year you can not be sure of the early hatches without a large incubator capacity. It is very much more desirable to have the chicks as nearly of one age as possible, as it will save untold worry and annoyance. If anything is more nerve-racking and confining work than a batch of newly hatched chicks every three weeks all through the spring, I have yet to find it.

You have in mind the size flock you want the following year, say 200 pullets to add to your yearling hens, so you need to buy 500 day old chicks. With good luck these would probably include about 200 pullets or perhaps a few more. Hatches as a rule will run about half cockerels. There would be a loss of some through natural causes—10 per cent. or more—the number varying according to hardiness of the chicks themselves, your care in bringing them up, and possible accidents.

You will find that in addition to possible carelessness, rats, cats, hawks, etc., love little chickens, so it will devolve upon you as foster mother to watch the flock with a mother's care. We will figure that you get your first lot of 166 on March 15 before which time you have ready two heated outdoor brooders (to be described later) which will take care of these 166 until about three weeks old. At that time have the second lot of day old chicks arrive, shifting the first lot to two colony houses (shown later) with heated hovers practically the same as the first as to heat, but larger in size. Then three weeks later still have the third lot of chicks arrive, shifting the second lot to two more of the colony houses, taking the heated hovers from the second lot which at six weeks of age and by April 26 should be safe without heat. The last lot could be moved from the heated brooder when three weeks old into two more colony houses, using the hovers from the second lot.

Should you desire to try for more than 200 pullets the first year, by all means increase each lot of day old chicks and add the extra brooder and colony houses rather than continue later in the season with more lots.

During the summer you will know how many pullets you have to winter and will add to the laying house, or winter quarters accordingly. We will, however, continue our estimates on the basis of the 200 pullets, approximating from actual expense the capital required and your results, figuring to carry the 200 yearling hens with their second year rather than to sell them, until the flock is larger.

ITEMS OF COST OF 200 HEN FLOCK

Cost of 200 pullets ready to lay	$200
Cost of housing in 40x15 house	150
500 day old chicks, 3 lots	75
2 heated outdoor brooders	30
2 commercial hovers	15
6 6x6 ft. colony houses, home-made	72
Water pans, feed troughs, etc. for 8 houses	8
40x15 addition to laying house to winter new pullets	150
Feed 200 hens 8 months, 10 cents a hen a month	160
Feed 200 pullets 6 months average	80
Feed 200 cockerels 4 months average	60
	$1,000

Approximate Receipts based upon above items

13,000 eggs March to July at 2½ cents	$325.00
2,100 eggs August to October at 3.2 cents	74.50
200 2 to 2½ pound broiler cockerels at 25 cents a ℔	115.00
50 same slow growing at 17 cents ℔	20.00
	$534.50

Less miscellaneous expenses, express, cartage, etc. 34.50

Balance $500.00

Inventory

200 hens and 200 pullets	$400.00
2 heater brooders	30.00
2 heated hovers	15.00
80x15 laying house	300.00
Pans, hoppers and troughs	8.00
6 6x6 colony houses	72.00
	$825.00

Total $1,325.00

This profit of $325.00 is low rather than high, as the 13,000 eggs might easily average 3 cents instead of 2½ and thus swell the profit to $390.00 or nearly $2.00 a hen. These estimates are not imaginary, or altered in any way from those on my books.

To apply previous remarks, remember you have had no living aside from the amount shown here, so you will understand why the first year is a trying one, and why you should know these facts before you begin. You are now in a position, however, to make a fair profit the following year as by this time you have a fair knowledge of the business from all sides, and having spent a healthful year outdoors, should feel ready to conquer anything. This should bring you success.

If you bought the 244-egg incubator at $32.00 the first year, you would by the end of the season understand running these machines and the results would have paid for it. Therefore, this item is not included in the estimates. You would also be prepared the following year to breed from your own pullets but should you do so it would be advisable to get two year old roosters of some other poultryman, for it is generally conceded that pullets and two year old roosters or vice versa make stronger chickens than do pullets and cockerels as parents. These estimates, as you will note, cover only eight months, April to December, as by that time the pullets will begin to lay and new calculations be made necessary.

Any directions given here for running an incubator would be superfluous as the manufacturers of incubators send out full instructions for managing them and each manufacturer knows the good points of his machine better than anyone else.

CHAPTER VI

POULTRY HOUSES RECOMMENDED

Following the plan laid out for starting the business, the house to be built first and ready for the 200 yearling hens or pullets by March 1, would be the laying house 40x15, which would also be their winter quarters. This house is really the most important of any since from the hens in it you would get most of your revenue for the year. In looking over pictures of egg farms you will find the majority of the flocks are kept in houses sometimes 200 feet long and housing 2,000 birds.

SMALL VERSUS LARGE HOUSES

If the majority of successful poultrymen use the long continuous house in preference to small detached houses, it certainly must be for good and sufficient reasons. The principal one is that it is more economical both in construction and general maintenance in the same way that apartment houses are cheaper than single detached houses. Another rea-

Front view of part of 128 foot laying house.

POULTRY HOUSES RECOMMENDED

son is that the owners are enabled to get an open front fresh air scratching house which would be impracticable in small houses.

There is no question that 1,000 hens divided into flocks of 25 with the proper house and space, under proper management—would outlay the same 1,000 hens if bunched together in one big house. For this reason you must not compare the records you may see of the man whose 25 hens in his backyard have done remarkably well, with the results you hope to get from your 500 or 1,000 kept together, because there are several circumstances that contribute to make a difference. The small flock, for instance, would have all the table scraps. This variety of diet makes the hens wild with delight, but the large flock would not be able to find the same quantity if put in their feed. The small lot would be more likely to get every attention in the way of feed, water and green stuff, where the large flock might be to some extent neglected. Again, unless you are trap nesting your layers, it would be much easier to weed out the drones of a bunch of 25 than from the large flock. Disease also could be more easily controlled in a small flock than in a large one. It is a common occurrence for all of a flock of 10 or 12 to lay in one day but not for all of a flock of 500.

Allowing that the results are better from small lots than when housed together in great numbers, you must stop to consider how you could properly care for 500 chickens if they were housed in 20 separate coops or 1,000 in 40 coops. The man who has tried the 3x6 house, housing 10 or 12 hens, certainly would feel discouraged as to ever building up a flock of 1,000 housed the same way, as the nuisance of feeding, watering and cleaning the houses would be so great, and the head bumps so many that he couldn't stand the strain. The cost of building is also much less with the large flocks, as is readily understood.

Long laying houses are all practically the same, 12 to 16 feet wide and any length desired. The one mentioned previously here, to house the 200, was 40x15 feet to be added to as required. Each additional 200 fowls will not need the extra 40 feet. My 120 foot house cares for 800 fowls, each one of the 800 having 1,800 square feet to roam around. The 200 in the 40 foot house would have only 600 square feet, so the large flock in one way would be better housed than the small

POULTRY HOUSES RECOMMENDED

Laying House. Interior View.

flock. Fowls should not be crowded to have less than 10 cubic feet of air space.

Compare the advantages of this house with the small ones aside from the cost. The cleaning of the droppings from the roosting board takes only a few minutes because there is plenty of head room and room to use the hoe, and the droppings are carried to a chute that dumps them in an outside manure shed; all in less time than you could clean three or perhaps two small houses. The disinfecting and whitewashing can be done in the same comparatively shorter time, with a sprayer and plenty of room in which to work. Feeding and watering, which with a number of houses such as you would have in the spring in raising little chicks, is a very considerable item of work, but in the big house is a very easy matter.

Best of all is the fact that all this work can be done, practically without going outside the house, but inside under cover from wind, rain, snow and sun. This one feature alone offsets any advantages the small house may have, as there are many days during the year when outside work is both unpleasant and unhealthy.

CAUTION IN HOUSE BUILDING

Supposing now you are going to build this long house 40x15, adding to it as your flock increases, there is one point to settle before beginning to build; namely, as to whether the house shall be built on the ground or raised up 3 to 5 feet. If your location is such that you have ample space and plenty of shade, then the house built on the ground may be preferable; provided you can arrange the floor so as to prevent dampness and do away with rats. The hens will find their way out and in a little better if on the ground than if raised very high. But if space is any object then by building above ground you add to your yard all the space that would otherwise be used by the foundation.

Another great advantage of the house being built 4 feet above ground would be that in summer when the hens are panting with the heat, they would have the coolest, shadiest spot possible under the house, dusting themselves, out of the sun, and in rainy weather, a dry spot which they would also appreciate. The last advantage to mention would be your freedom from rats or other animals; for rats will never bother you in a house raised off the ground sufficiently to see under it. Following is a description of this house which any mechanic can build without any further plans.

DESCRIPTION OF POULTRY HOUSE

Width, 15 feet inside, 16 feet outside; height, 8 feet front, 5½ feet rear; length, 40 feet or longer; studding, 2x4's, roof rafters 2x6; floor timbers 2x6 or 8; sills 4x6; roof, floor and sides 8 inch N. C. T. G. flooring, all papered on the outside with best quality

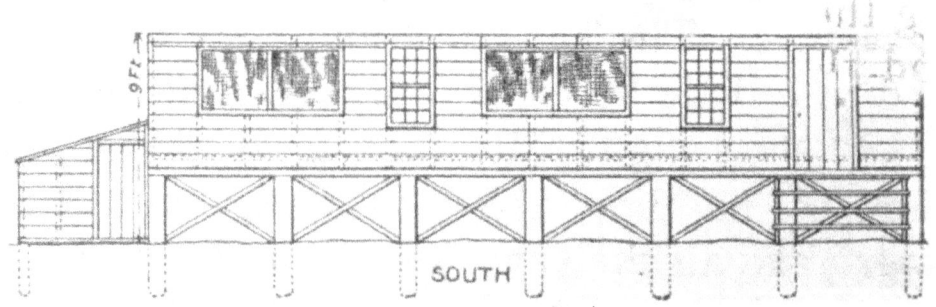

Detail of front of a 47 foot section of laying house.

of roofing paper, with joints cemented. Cover the roof with the best paper as it is poor economy to use cheap paper there, as you will lose more than the first saving in repairs and sick chickens which a leaky, damp house is sure to cause.

There are no windows in the sides or the rear, but every 40 feet of the front has a double sash window for light on dark, rainy days when the curtain windows must be

POULTRY HOUSES RECOMMENDED 51

closed. These curtain windows are 9x3½ feet. One of them is made in two sections, in each 40 feet of the house, covered with light duck or bed ticking. These windows swing in and up to the ceiling. They are kept open during the day and closed at night, air passing through the duck when shut down, giving the necessary ventilation. Poultry netting is tacked on the outside over the openings to prevent the birds flying out. When these curtain windows are raised during the day you have really an open air house and the sunlight reaches nearly every part if the house is built facing the south.

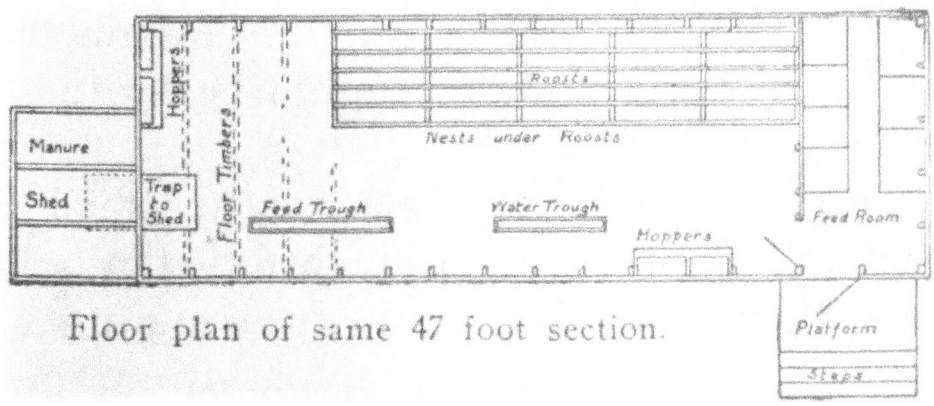

Floor plan of same 47 foot section.

A roosting and dropping board is built along the rear of the house 3 feet from the floor, 6 feet wide, made of T. G. boards running across, the way the hoe will move when cleaning. Five roosts, set 9 inches up from the board on 2x4 legs, run lengthwise of the dropping board. For very cold nights I have curtains which let down to this dropping board enclosing the hens in this small space and making the danger of frozen combs very remote.

Under the front edge of the dropping board is a line of nests made of two 8 inch boards, the lower one nailed tight but the upper one hinged to let down for gathering eggs and cleaning nests. When the upper board is hooked up it presents a solid surface from the front, darkening the nests, entrance to which is from the rear. Dark nests are preferable to exposed ones as with them hens are less likely to develop the egg-eating habit.

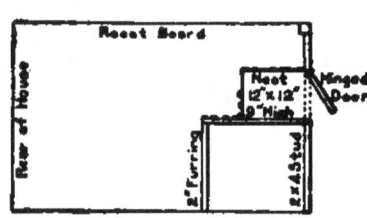
Detail of nests.

Feed hoppers for dry mash, oyster shells, grit, etc., are placed along the walls—but not under the windows where the feed might get wet—on platforms about a foot from the floor. The feed and water troughs are V-shaped, both raised from the floor and with boards over their tops so the birds can reach the feed and the water, but cannot step into them.

The floor is covered with 6 or 8 inches of straw thrown in from the bale. It is soon

POULTRY HOUSES RECOMMENDED

scratched to fine pieces. In this the grain is fed, making the birds work for the food. Some writers say this litter can remain until spring, but cleaning out and renewing at least once is preferable, as it becomes very dusty and dirty before the winter is over. One objection I have found to straw as litter is that straw seems to draw dampness from the air thus becoming foul and musty. I am now substituting planer shavings, which I find very satisfactory.

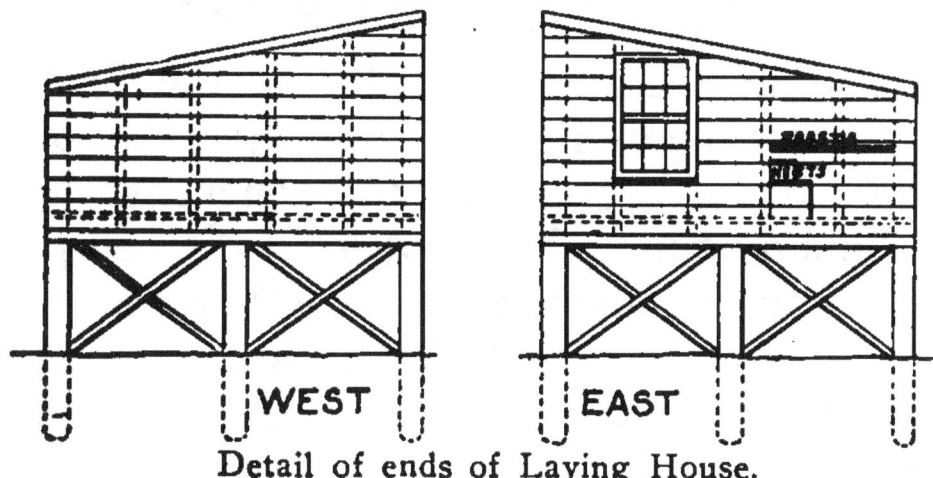

Detail of ends of Laying House.

Outside one end of the house a covered manure shed is built, connected from the inside with a chute covered with a trap door through which the droppings are thrown after each cleaning. By spring a good many dollars' worth of manure can thus be collected. All farmers are glad to get such fertilizer.

The house should face south if possible. If you have the space it would be advisable to place it in the middle of a plot with 100 feet on each side, front and rear. In such case you could plant oats or rye and after one side was cleaned up, turn the chickens on the other patch and replant the first one, thus having a continuous lot of green food. Rye and oats grow very fast. A couple of weeks will renew the growth from the browsed stumps.

HEATED BROODER HOUSE

The next house in order is the heated brooder. There may be others just as good, but this one is good and the general style is the main point. This brooder is divided into two sections, one heated, the other not. At least a day ahead of the arrival of chicks, the lamp should be lighted, the burner adjusted and the heat regulated by means of the thermometer. The proper temperature is 95 degrees. When that is reached the thermostat should be adjusted to hold the temperature. When you are convinced that it is right the chicks may be placed under the hover, the heat from which comes from above, imitating to an extent, the brooding hen. Directions come with the houses, but care must be used not to allow the temperature to rise too high, for while the apparatus is self-regulating to a degree, still it has happened that chickens have been overheated, which danger is greater than that they will not get enough heat.

Brooder raised chicks depend upon you to supply some of the attentions that the mother hen supplies. It is best that only one person attend to any heated brooder or incubator, so as to avoid the duplicating of any detail and the missing of any other.

A partition divides the heated apartment from the non-heated or exercising room, entrance to which is through a little door. When opened it does not take the little fellows long to discover the other room. They soon run back and forth from the cool to the warm part, but they should be watched at first for fear some will forget the way back and so get cold. Chaffy hay seed should be scattered on the floor of both apartments to the depth of a couple of inches, keeping the floor clean and inducing exercise. Little chicks can be kept safely in these houses in very cold weather. In such weather this seems the only practical way of brooding them.

Some poultrymen advocate a fireless brooder; i. e., one without artificial heat They depend on confining the heat which the chicks give off from their bodies to keep up the warmth. My experience is that little chickens brooded in any hover without artificial heat crowd one another. Crowding must be avoided always, or a loss will result, even when the chicks are large.

Home-made fireless brooder.

Without heat they always crowd, looking for the warmest spot in the center of the bunch. The continual pushing of the outside ones often results in the middle ones being smothered.

Same, showing corners rounded.

The mother hen in brooding separates the chicks with her legs and her feathers, besides supplying them with heat.

If you ever use any brooding device without heat, make sure the corners are rounded out, as it is to the corners, always, that the little fellows go first when sleepy. Rounded corners will dissipate the danger to some extent. Heated brooders will accommodate 50 to 100 chicks depending upon size. With the small loss one is almost certain to have, they will care for the little flock until the next batch arrives, when the chicks can be

Simple fireless brooder.

moved to heated colony houses. You will need water pan, feed trough and little grit and charcoal hopper for this brooder. All of these must be kept scrupulously clean. Should the season be a mild one, this house could be omitted and the chicks brought up from the day of hatching in the colony house, description of which follows.

BROODER AND COLONY HOUSE

I built my brooder and colony houses 6x6 feet square, 6 feet high in front, 4½ feet high in the rear, with sloping roof. They are made of 2x4 studding and 8 inch N. C. T. G. flooring. The sides are papered and the roof covered with roofing paper which is satisfac-

Colony Brooder House showing hover.

tory provided enough nails or battens are used. The house has a floor laid on top of the 2x4 sills, which prevents dampness and keeps out all animals. To make sure of freedom from rats, a 2x4 run on edge on the bottom of two sides raises the house 4 inches off the ground and leaves an open space. If any

POULTRY HOUSES RECOMMENDED 59

house like this rests directly on the ground, with a space under the floor enclosed on all four sides, it is always an invitation to rats to make a nest; never when two sides are open.

One half of the front is hinged to swing out and up, to act as a ventilator, as a shade from sun and a protection from rain driving in during the day. The opening made by

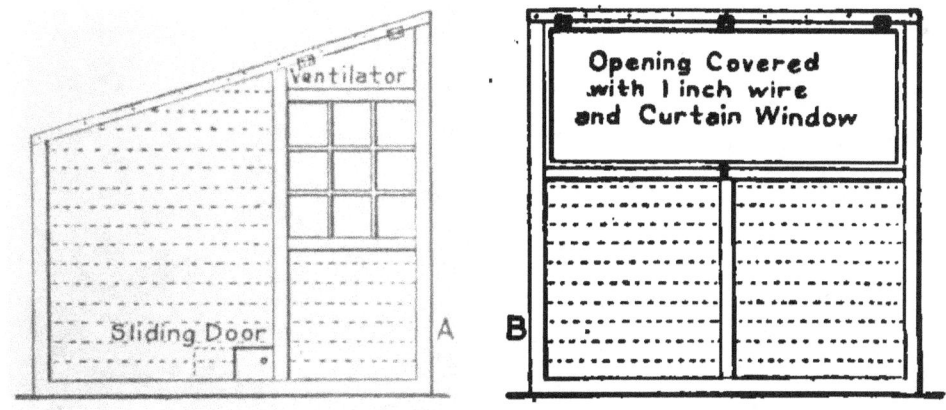

Details 6x6 Colony Brooder House.

this window is covered with 1-inch wire netting to keep birds in if necessary, or to keep animals out on warm nights, when this window is hinged up. A door is set in one side, and a small glass window in the opposite side for light on stormy days when the front window is closed. A small sliding door at the floor level allows the chickens to go out and in. This little door must be shut at night.

A commercial hover is placed in the center of the floor. After being warmed up and after you have put in 2 or 3 inches of chaffy hay and seed and sand for scratching litter the house will be ready to receive the little chicks, three weeks or so old from the outdoor brooder. A feed trough, a water pan, and a hopper should be in each house. Thus the chicks are housed for any kind of weather.

Commercial hovers used in 6x6 colony houses. These may also be substituted for the hovers of more complicated construction in the brooder house.

Should the weather be cold or rainy, the front window can be closed. The glass window will furnish light so the chicks can work in the litter for feed, and there will be plenty of space to accommodate them. If on the

POULTRY HOUSES RECOMMENDED

other hand, the weather should be fine and warm, the little fellows can run outside. The sooner they get out on the ground the better. Should the weather be very cold so the chicks cannot be kept warm enough, a temporary ceiling of unbleached muslin could be set about 2 feet or so from the floor, to reduce the cubic space to be heated.

As the weather becomes warmer reduce the heat. By the time the chicks are six weeks old the hovers can be removed to the other houses for the next lot of three weeks old chicks. At this time, when the heat is taken away I have always had some trouble, as the little birds have been accustomed to heat at night and even if not necessary, they are likely to crowd. They must, therefore, be watched.

Water Pan Feed Trough and Feed Hopper required in all colony houses.

A week before the hover is taken away put in roosts 8 or 9 inches from the floor. Doubtless some chicks will begin using them, even if only during the day. When the hover is removed, they may start right in roosting; but if they do not of their own accord, each night place some of them anyway on the roosts. Once you get the six weeks' old chickens roosting, you can almost count on saving them all. In the plant of

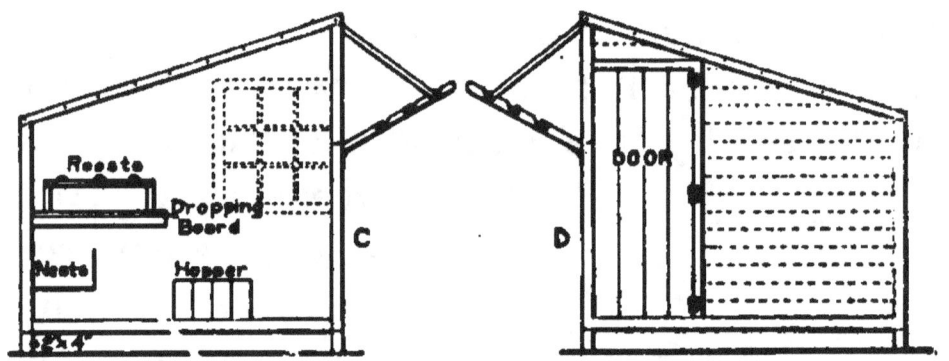

Showing Brooder House converted to Roosting House for fowls.

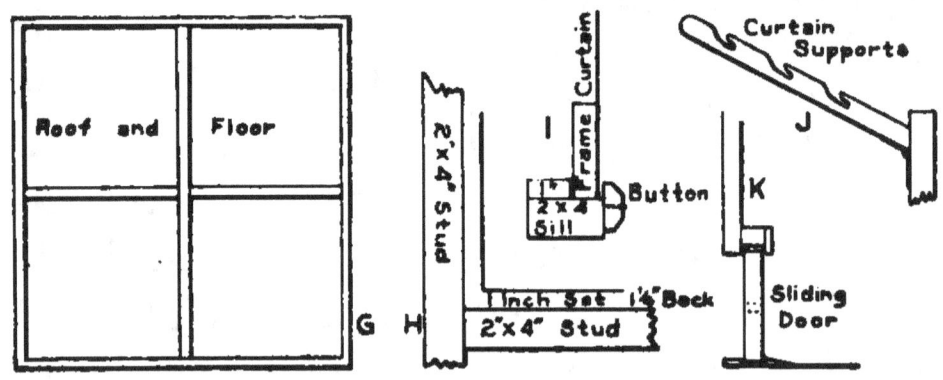

Details floor, roof, and window.

the novice who is experimenting with little brooding devices of his own make, more deaths are caused by crowding than from any other cause.

POULTRY HOUSES RECOMMENDED 63

As the chickens get near the broiler size, the cockerels which are readily distinguished from the pullets if your breed is the Leghorn, can be removed to a separate pen for special fattening. The additional space thus made will allow for the extra growth so the pullets can be left in this house they are accustomed to, until final removal to their winter quarters. The houses should be set 40 to 50 feet apart so each flock will

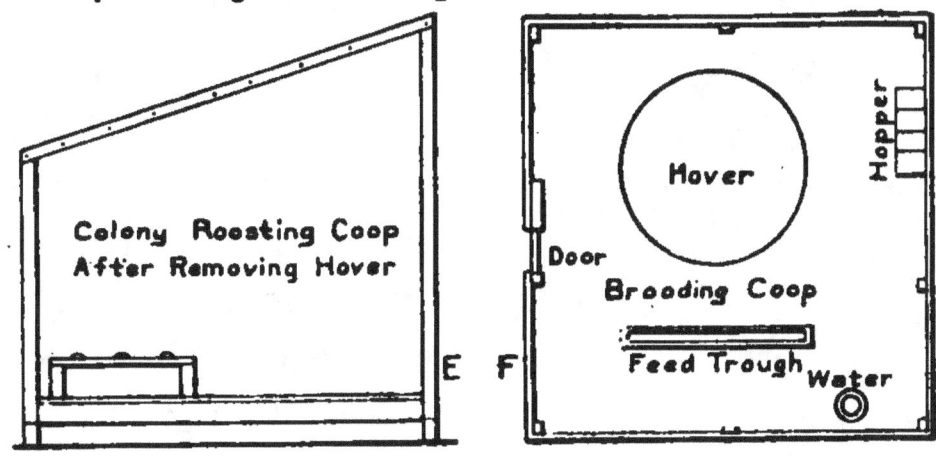

learn to know its own house. If when first moved from the heated brooder to the colony house, a wire run is used to keep them by their house. They will become accustomed to it and always use it.

When used as a brooder house the corners of the house should be rounded out with Beaver board, tin or other material to prevent chicks crowding into the square corners.

HEN BROODER COOP

Perhaps you will want to use the hen to hatch some chicks, allowing her to brood them. In this case the little house shown herewith will be found the best possible style. It is two feet long, 18 inches wide, 20 inches high in front, and 16 inches high in the rear, with a 12 inch overhang roof all the way round. It is made in two parts as shown, the lower part of twelve inch boards, sides and rear, and the upper part 8 inches in front and 4 inches rear, the side pieces being ripped from one 12 inch board with one ripping. The top is exactly the same size in length and width as the bottom upon which it fits, held in place by four little posts projecting up 3 inches from the bottom to which they are screwed. The advantage of this arrangement is that the house can be cleaned easily, simply by removing the top.

Hen Brooder Coop, wire door in place for night; also showing how coop is split in two to facilitate cleaning.

POULTRY HOUSES RECOMMENDED

Nail a floor on the bottom and then a 2 inch or smaller strip on the bottom of the floor at the rear to raise the floor of the house to slope toward the front and tend to drain off any water that might drive in the door.

Two little doors are then fitted, one covered with 1 inch wire mesh to use at night to keep

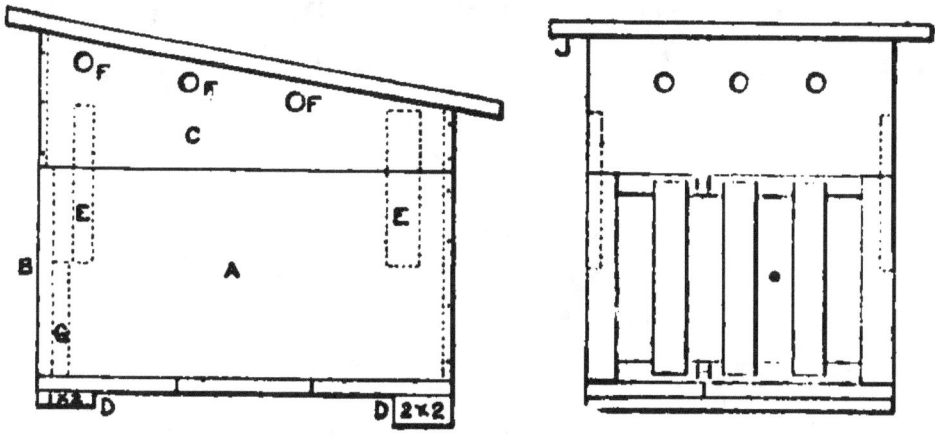

Details of Hen Brooder Coop.

the little chickens in and rats and cats out, the other with slats to keep the mother hen in, but allow the little chicks to run out and in. Cover the top with roofing paper and paint. All houses should be painted, not only to preserve the wood, but for the sake of appearance.

SMALL COLONY HOUSE

For 25 or so chickens just removed from the heated brooder at three or four weeks, the small colony house is very satisfactory one. It is 6 feet long, 2 feet wide, 2 feet high in front and 18 inches high in the rear, made of 3 inch ceiling boards T and G, the roof papered over the hover. The hover is formed by a partition across the house 2 feet from the end. This partition runs up to the roof at the rear, but is 6 inches from the roof in the front, thus giving ventilation. The space between the top of the partition and the roof is covered with wire

Small Colony House.
No heat. Notice tin rounding out corners.

for protection from animals at night.

The hover has a floor 2 inches off the ground. The 2 feet comprising the hover is sealed solid on the side and the rear, but the front is made as a door, held in by buttons so that as the weather becomes warmer, the door can be lowered 2 inches still being held in by the buttons, but giving a ventilating opening at the top. The four corners of the hover are rounded off with tin When the floor is covered with hay seed, it makes a cozy little hover. The partition has a little sliding door to shut the chickens in at night, making an animal proof house.

Roosts can be put in the other 4 foot end. Later the little chicks may be shut out of the hover and made to roost, this end, of course, being wired up with an additional hinged door to allow the chickens to run in and out.

BROODER AND INCUBATOR HOUSE

If you desire to build up a large flock by incubating the eggs from your own stock, it will be necessary for you to have some form of incubator cellar and brooder house.

The photograph and drawings show a very convenient and economical form of building for these purposes. The incubator cellar will accomodate 4 good sized incubators totaling 900 egg capacity, and is partly below ground.

The hovers and run will in turn take care of the little chicks hatched in these incubators until they are three weeks old when they can be moved to the outdoor Colony brooder houses.

On a fully equipped plant this building would be heated by hot water, but this system is quite expensive so I arranged to heat the cellar with a small stove and the hovers with oil lamps. Commercial Hovers could be used in places of the ones shown.

Brooder House, 16x42, half underground.

During the past year or so another method of brooding chicks in large numbers has been evolved by which flocks of from three to ten hundred can be raised under one brooder.

These brooders are built after the general idea of the commercial hovers shown, only they are greatly enlarged and the heat is sup-

POULTRY HOUSES RECOMMENDED

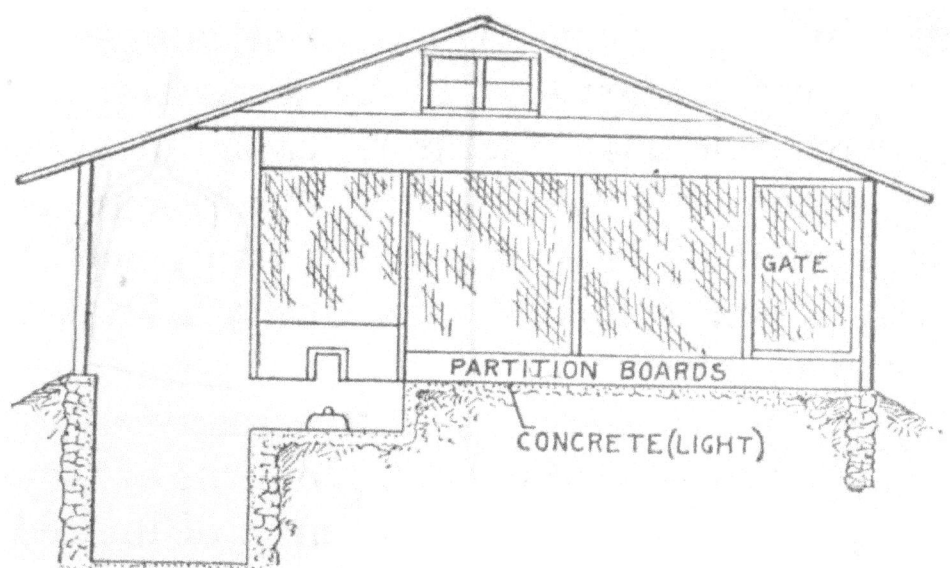

Detail of Brooder House.

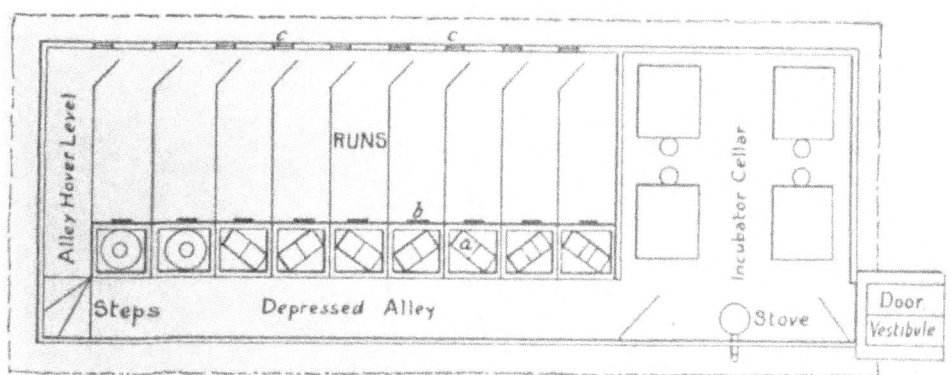

Floor plan, brooder house and incubator cellar.

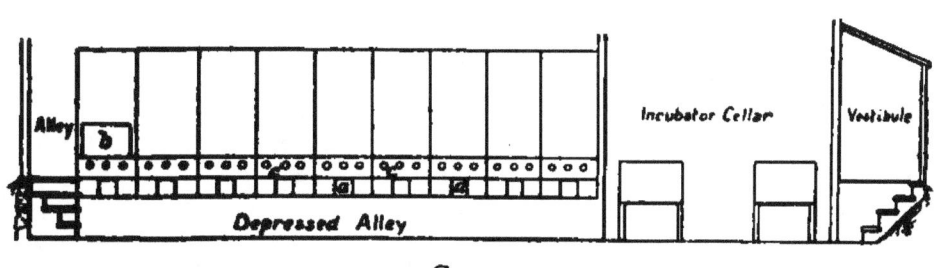

Same.

plied by a coal stove in place of oil lamps.

If these Mammoth Colony Brooders will do what the manufacturers claim, it will surely simplify the brooding of chickens in large numbers, a problem that has troubled many a poultrymen who had only a limited capital to invest.

The poultry business is like many other lines of work—what may be problems of very annoying proportions to the man of limited capital are easily overcome by the man who has sufficient means to install all the modern and best appliances.

CHAPTER VII

BREEDING AND HATCHING

The natural hatching time is in the spring, from the middle of March forward according to the coldness of the climate. Any hatching done before that time is very likely to be disappointing. One may read of very early hatches in January and February but if they are encouraging in percentage he can rest assured that the conditions governing the fertility of the eggs have been the best. The novice will be better off not to try for the very early hatches unless he can meet those conditions.

In the first place it is not Nature's time. That fact probably has everything to do with the other conditions. Most fowls have had about all they could do to get through the winter with the cold weather, the confinement and lack of some natural conditions. They have not had as good air, as much sunshine, or the food and the exercise that they get when on range, consequently they are not in good condition to produce strong, fertile eggs. Again the eggs are likely to be chilled so that altogether the chances are for weak germs which may start and so raise one's hopes, but finally result in very poor hatches. Because of these difficulties do not set eggs gathered from other people, but if determined to hatch early, have your own breeding pen.

It is hard to say which method of penning fowls is the best for the fertility of the eggs,

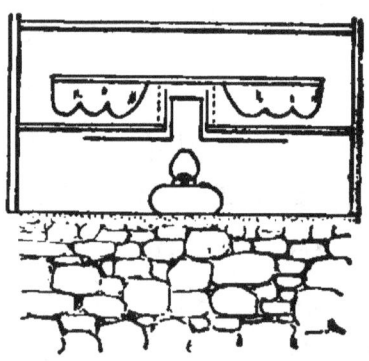

Hover construction in brooder house.

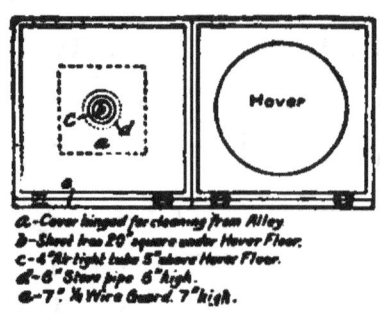

Top view of hover.

whether one cock with 10 to 15 hens penned up separately, or several males with a large number of females. I believe either way is satisfactory and that the results depend upon the quality of the fowls penned up. If two or more males are to be in the flock, the size of the enclosure must be considerably increased otherwise one male will interfere with the proper work of the others.

Most farmers, of course, work on the plan of a large flock with plenty males, but also plenty of range. During the natural breeding season the eggs from farmers' hens are very fertile and strong. It is a common occurrence for them to have every egg hatch, not as the incubator would say "every fertile egg," but every egg set. This shows that hens must have range or plenty of exercise and good air to lay fertile eggs.

Eggs from hens too fat, too poor, or from hens that have been forced for winter laying do not hatch well.

Eggs to be used for hatching during cold weather must be gathered often and must be taken care of so they do not become chilled.

Fresh eggs hatch sooner than eggs a week older.

If eggs are to be kept any considerable time to use for hatching, they should be turned often to prevent the germ and the yolk settling.

Hatching eggs received from any distant point should not be set for at least 24 hours so the various parts may reassemble.

The sex of the chicken that will hatch cannot be determined by any examination of the egg.

Hatches as a rule produce about as many cockerels as pullets, this probably being controlled by the vitality of the male.

The male is the most important item in the breeding pen as he effects every egg, while each particular hen effects only the eggs she lays. Whether eggs are hatched with an incubator or by a hen, the results are mostly dependent upon the eggs to be hatched as neither hen nor incubator can hatch poor eggs.

Commercial case for holding and turning eggs while waiting for incubation.

Care should be used in the selection of hatching eggs. These should be of medium size, not very small nor exceptionally large, with smooth shells without ridges, free from cracks and reasonably clean.

Every bird to be used as a breeder, either male or female should be selected with the idea of using only the best as breeders of future flock.

Find out the points or qualifications needed to make perfect birds of the breed you intend to handle, and select each bird with the points coming nearest those qualifications.

Some writers advise mating pullets to two year old roosters or vice versa but it seems to me the main point would be that the breeders should be strong, healthy birds. I believe that fully developed pullets after having laid their first small eggs, if bred to fully developed cockerels would produce as strong healthy chickens as if the ages of the breeders, were different.

CHAPTER VIII

INCUBATING

I have advised buying day old chicks the first year as a more sure way of ending the season with a certain number of pullets of a profitable age, but the following year it would be cheaper to hatch your own chickens or at least part of them. A few remarks on the work of caring for incubators may be helpful.

The old way of hatching with hens is nearly an out-of-date method for any poultryman whose plant numbers several hundred and who is running that plant for profit. This is so for several reasons; viz.:

First, hens to be most profitable must be brought to laying age some time in the early winter. As pullets raised by the ordinary poultryman take six or seven months to reach this laying age, this requires that the chickens must be hatched in the latitude of central New York State by May 1 at the latest.

INCUBATING

Most hens do not get broody before that date, therefore a poultryman can not depend on getting early hatches. Hens are not like incubators in that you can set them when you want, but you must wait upon their own sweet will, which is a poor way to base your calculations. Second, if your flock is of Leghorn hens you can not calculate at all upon having hatchy hens.

Third, if your breed is one known as a broody kind and you desire to raise several hundred little chicks, even disregarding the question of whether they should be early hatched or not, the care and annoyance of a large number of hatching hens would be very considerable. Fourth, it is very much more expensive to hatch with hens than with an incubator. Supposing you have a 244 egg incubator we will compare the expense:

The eggs would amount to the same in either case. It would require 16 hens to set on the 244 eggs, and these hens would set for 21 days and would consume some 42 days more brooding the chickens after hatching —a total of 63 days. These hens could have been broken of their broodiness in 10 days at the most, when they would be laying again almost every day or 53 days, or say 40 days.

Sixteen hens laying 40 eggs each for that period would be 640 eggs. At 2½ cents this would amount to $14—fourteen dollars worth of eggs as against the cost of running an incubator and brooders the same length of time, which would not amount to more than $3.75 for the oil. The time spent by the poultryman would probably be about the same in each case, as hens have to be watched and cared for while hatching as also while brooding their chicks.

On the other hand, my experience has been that hatching hens will bring off a larger percentage of chicks than an incubator will. I have run incubators for several years but never yet have I had a 100% hatch of the fertile eggs or anywhere very near 100%. I have had the best machines and have followed the directions as nearly as I could understand them, but still the hens would beat my incubator. The greatest trouble is dead chickens in the shell, chickens well developed but without strength enough to come out of the shell.

I have read the explanations usually given —the germs were not strong, the fault was in the eggs—but I am positive that reason alone was not correct because at the same

INCUBATING

time the incubator was set, hens were also set with eggs just as fresh and coming from the same breeding pens. Why the results should be so I do not know, except that the natural instinct tells the hen to do some things under the varying conditions of temperature and atmosphere that are not apparent to me.

Manufacturers of some incubators claim the machines do not need supplied moisture, while others have a provision for additional moisture. I believe the hen has the instinct and the power to supply moisture as needed, as well as to know when to leave her eggs and when to return.

The directions for airing the eggs supplied with my machine say that "some operators do not air the eggs any more than the time it takes to turn them, while others air them a considerable time, and that the operator must use his own judgment." From results sometimes it would appear that our judgment does not compare with that of the hen. However, when all is said and done, I will stick to the incubator in preference to the hen.

In the first place, buy a good incubator, set it up, making sure it is level, otherwise the heat may be more at one end than at another. It should be in a cool, well ventilated place, partly underground if possible, so the air will not be very dry. Test the thermometer with another of known accuracy such as a physician's, and if your thermometer varies one way or the other, note it. Light the lamp, turn the wick to a medium height, and let it burn for a day or until the thermometer registers 103 degrees, at which point set your thermostat. When you are sure this temperature is maintained you can set the eggs previously selected.

Now, follow out the directions sent with the machine and pray for a good hatch, meanwhile being careful to do the following: Look at your thermometer each morning before opening the machine to turn the eggs. If the temperature is not right, make it so. Fill the lamp every morning and trim the wick. Turn the eggs and an hour after again look at thermometer to see whether the lamp has been turned too high or too low. Turn the eggs again at night. At other times leave the incubator alone. Be careful not to allow anything to set on top of the incubator where it might be pushed against the thermostat rod and thus result in the temperature running up with disastrous results.

INCUBATING

Let one person attend to the incubator, otherwise some detail will be neglected or done twice. Test the eggs for infertile ones any time after the seventh day, either with an egg tester or by holding the egg between the thumb and first fingers, making a shade of the rest of the hand. At seven days the infertile eggs are readily recognizable because they are perfectly clear. They should be removed, boiled and fed to the chickens.

When a hatch is coming off, the incubator should be kept closed. Any chickens worth hatching will break out themselves. It is not often that the chickens you help out of the shell are worth saving. A good hatch should start on the twentieth day and be over the twenty-first day. A late hatch usually results in a number of cripples or weak chickens that usually die before many days.

Leave the chicks in the incubator for 24 hours or until they are thoroughly dry. Meanwhile get your hover or brooder in readiness. After the hatch is over remove the trays, brush out all the dust and clean off all stains ready for another setting.

Once an incubator thermostat is adjusted to 103 degrees there is very little difficulty in keeping the proper temperature, the times giving the most trouble being right after the lamp is refilled and trimmed. At such time as there may be a decided drop in the outside temperature, an incubator will keep up to 103 degrees when the outside air is below the freezing point. The most satisfactory results are attained however when the room in which the machine is set has a temperature of 65 degrees.

The beginner will have enough to do during his first year without having to worry about incubation and wondering and speculating upon the results of his hatches, which might be very poor indeed, setting him back in his calculations and disappointing him at the outset; but if the day old chicks purchased are of the breed and strain that he would want if he bred from fowls or hatched from bought eggs, he would have just as good foundation stock for the future and could make his plans as to how many he wanted and when, and would get them. This would be a doubtful question if he hatched them himself.

If he expected to hatch from his own 200 hens, the number as here planned, his receipts from eggs would be materially reduced and he would have to pay $300 to $400 for his hens instead of $200, as the hens I have figured on here were not for breeding but just a temporary utility egg-laying flock. If he bought the eggs to hatch he would have to pay 6 to 8 cents for the same strain as his 15 cent day old chicks, so that before he got through and had the 500 estimated he would find that his flock had cost him at least as much as the 15 cents apiece.

Incubators in the hands of novices don't hatch every egg bought nor every fertile egg. So if you hatch 50% to 60% of the bought eggs you will be doing well. Figure it up and add your oil and see how close the figures would be to 15 cents.

To breed properly from your own hens requires also more study and knowledge than the average novice would be able to master to enable him to start right in to raise his own flock and keep that flock up to the standard he wants. It means study, watchfulness and a knowledge of his hens that a beginner would not have; so but few words have been said as to brooding because it seems better that the beginner should follow the other plan and take the first season to profit by his experiences and together with his reading use the knowledge the following year.

The plan contemplated is a short cut to a large flock in a short time, giving the owner time to get his experience, however, before he has much money invested. It is not a short cut by new fangled methods, but the one I selected after experimenting and approved by the majority of poultrymen. After having tried other less satisfactory methods, I settled on this. By it the novice can easily do in two years what has taken me three years to accomplish. You may very easily exceed all my estimates, for my flock has been a very ordinary one, but it has had good feed and attention.

Various ways have been noted throughout the book of increasing profits over mine. The whole object of this book has been, not to try to induce you to embark in the business, but to put everything in the business, as far as has come to my mind, in as clear a light as possible and in a conservative way, leaving it for you to judge whether or not you could make a living from chickens.

CHAPTER IX

POINTERS ON HEN HATCHING

In setting a hen it is generally conceded that a piece of sod placed under the straw of the nest aids in getting a good hatch. The nest should be large enough for the hen to turn around and manipulate the eggs, should have hay or straw enough that the eggs will not touch the boards or become chilled.

An odd number of eggs should be set—13 or 15—as the eggs will round up in the nest better if an even number is used.

The hen should be set on china eggs first to make sure she will stick to her job, and should be watched at first as many a hen has left her nice little nest of eggs and selected one with two or three china eggs in preference!

Hatching hens should be kept in a quiet place, should have water, feed and a dust bath. Great care should be used to see that they are kept practically free from lice. They are often known to die on the nest, literally murdered by lice.

If you set three hens the same day, two will be able to care for the resulting hatch and the third may be broken up and started laying again.

Give the brooding hen and her chicks a roomy, dry house with some litter on the floor, but not too much while the chickens are very young. Keep the house clean. Lock or close up the coop at night to prevent enemies getting at the brood. Let them out during the day and it will be a sight to warm your heart as you watch her show her little ones what to eat. If you have a vegetable garden your heart will warm in another way as she shows them how to scratch.

Not every good setting hen makes a good mother, as some hens treat their broods in a far from motherly manner, picking them and otherwise ill-treating them, so that losses among hen brooder flocks are sometimes heavy.

If you have more than one hen hatching in the same coop you will have to watch them to prevent one hen from leaving her nest and setting on the nest with another of the hens.

Setting hens may leave their eggs at any

POINTERS ON HEN HATCHING

time, even after they have been setting for several days, often during the last week. So it is advisable to have another broody hen ready to take the place of such a quitter, otherwise your eggs will be lost. Do not set any Leghorn hens, because as a general proposition they are poor setters, having a nervous disposition, which cannot be counted upon to keep them on the nest three weeks. Occasionally a Leghorn will lay a clutch of eggs in a stolen nest where, free from disturbance, she will bring off a brood, but if you wish to have good success, select a naturally broody breed rather than the Leghorns to set.

I will now ask a question that I cannot answer, but will bring forcibly to your mind the workings of Nature. I have stated here and all others say the same thing, viz., use fresh eggs for hatching for fresh eggs will hatch better and sooner than older eggs. How then does a hen which steals a nest lay a clutch of say 15 eggs, the first one perhaps 20 days older than the last one, and each one from the first to the last having been set on say half an hour a day for 1 to 19 days, while she has been laying each succeeding egg and some of which we would not call fresh—

how does she bring out a hatch of the whole number?

CHAPTER X

FEEDING CHICKENS OF VARIOUS AGES

For the first few days all poultrymen feed their chicks about the same way, no matter how much their methods may vary later. For 48 hours after hatching, the chicks should not be fed anything at all, as they get sufficient nourishment from the yolk of the eggs they absorbed just before hatching. The first day of feeding I give mine stale bread rolled up fine and oatmeal moistened with milk, but the milk should be squeezed out so the mixture is flaky rather than pasty. This should be fed until they have eaten all they want, four or five times a day at regular intervals. After they are satisfied the troughs should be removed and cleaned to prevent whatever is left from souring.

The most convenient way to feed is on a board about 3 inches wide with a lath nailed around the edges to project up about half an inch, the ends raised up about an inch above the feeding board and another lath set on top so the little chicks can reach under and get their feed but cannot step into it and pack it down or soil it.

Water or sweet milk is given in some kind of pan that will allow them to drink, but so arranged that they cannot get into it. From the very first day chickens will do just the things you don't want them to do. There are various styles of galvanized drinking founts. The smallest size made in the style shown in photo is very satisfactory.

Little chicks feeding at covered trough.

Perhaps the first time the chicks are fed most of them will not touch the feed because they are not used to the operation. But show a few of them what they are expected to do and from then on there will be a regular rush to feed and water. You should stay by the little fellows the first few times they are fed, to see that none of them fill up and then stand around and get chilled. After they appear to have

FEEDING CHICKENS

had enough, push them back under the hover, where they will stay until the next feeding.

The second day's feeding is practically the same. A little hard boiled egg may be added to the mash and a handful of chick feed scattered in the litter with some chick grit added The third day, reduce the soft feed to three feedings; the fourth day to two, and from then on either cut it out altogether or feed only once. Some poultrymen advocate omitting the mash feed. They rely entirely upon the chick feed; but chicks relish one feed a day of mash. Although it adds to

Covered water trough to prevent fowls befouling the water.

the attendant's trouble, still I give it. Chick grit, charcoal and bran should be kept before them all the time in little hoppers. They like bran. The main feed, however, should be the chick feed sprinkled in the litter on the floor. The deeper the litter the more exercise the chicks have to take to get their meals. Exercise is the great health-giver. The drinking founts and the feed troughs should be kept scrupulously clean. Stale feed and filthy water will cause sickness.

Keep the hovers at 95 degrees the first week and then gradually reduce as the chicks grow older, the reduction depending somewhat on the weather. The chicks must be kept warm. By watching them you can tell if they are comfortable. If they are not, make them so. They will not grow into strong, healthy chickens if continually chilled or overheated. They should be allowed to run out on the ground a little each day as soon as the weather will admit of their doing so; but a wire run should be put up to keep them close to the house as well as for protection. You will have to watch the first few times they run out to see that they know enough to run back into the warm house when they become cold.

Home made single feed hopper.

TEACH CHICKENS IN THE WAY THEY SHOULD GO

Every time chickens are allowed to do some new thing to which they are not previously accustomed, they must be shown what is expected of them, and forced to do it until it becomes a habit. Remember this all the way through your work and it will save you losses. You must not expect to open a door for the first time and let your

Colony House with wire run to accustom little chicks to their own house.

chickens out into a strange world and believe they will all return. The mother hen takes her brood around continually clucking and calling them to her. While you can't cluck, still you must supply some of the mother hen's attentions if you expect to save your chicks. Most of the losses of little chickens occur through carelessness. Disease takes some, but perhaps even the disease was caused by neglect.

Usually the first little chickens are a novelty and so receive every care and attention, but as more come, making more work and less novelty, carelessness is likely to creep in. Then chicks are lost. It is far better policy to hatch 100 chicks and raise 90 than to hatch 200 and through carelessness raise only 100 or so. They go fast when they commence to die; so be as careful of the last hatch as of the first, and your books will look much better than they will if you get careless toward the end.

The feeding of the fifth day is repeated daily until the chicks are six weeks old, but in addition a little green stuff, such as lettuce or sprouted oats or tender grass is added to the diet, sparingly at first. Beef scrap, brooder size, is also kept in the hoppers.

The plan outlined in this book calls for the removal of the chicks from the heated brooder house to the heated hover colony house at three weeks or so, but the change must be effected with as little change of temperature between the houses as possible, the colony houses having been previously warmed up to the proper temperature.

Plenty of hay seed or other suitable litter in which there has been considerable chick feed scattered must be on the floor.

REMOVE HOVERS AT SIX WEEKS

At six weeks of age the weather should be such that the hovers can be removed, the temperature having previously been steadily reduced, so there is not much change when the heat is removed. Then if you can get them roosting your troubles will be almost over. At this age scratch feed, or a larger size of chick feed is gradually given in place of the small chick feed, and a growing mash is fed dry in the hoppers. A wet mash may be fed once a day if you wish. It should consist of meal, bran middlings and a little beef scrap; but this is not necessary. Chickens of this age with a hen would not get anything but grains, grass, bugs, grit, etc. In growing any living thing, Nature's way is the best. The feeding suggested with water, grit, charcoal, always where they can get it, is continued until the chickens are removed to winter quarters, which should be about October 1 to 15.

To grow the sturdiest, healthiest laying stock requires that the pullets have plenty of exercise or range. Here is where the farmer with a large place for his chickens to run has the advantage over the man with more limited area. Hens do not require much range, in fact, do well with none, but pullets to grow properly, must have the space to roam and develop.

Home-made double feed hopper.

One great lack in the summer time is green food, especially on a small place, as a flock of growing chickens will clean up a limited yard of every piece of green stuff in sight, so when your ground has become bare, green stuff will have to be supplied. Sprouted oats and alfalfa will answer the purpose. Sprouted oats are grown by soaking oats overnight in hot water, then placing in frames and watering twice a day; but for a large flock of chickens to get green food from this source would require considerable space to care for it. Alfalfa suits me.

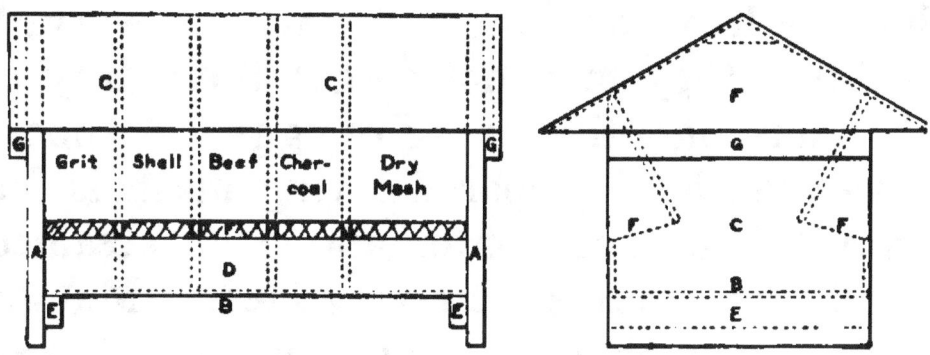

Details of double feed hopper. End detail of double feed hopper.

FORCE PULLETS FOR WINTER LAYING

When the pullets are put in their winter quarters they should be forced for laying, as laying eggs in quantities is what the business outlined in this book is built upon. Breeders who raise chickens to sell as foundation stock to others, or who want fertile eggs for hatching, probably do not care so much as to the quantity of eggs but must have a hardy flock of healthy natural birds which they would not force by feeding specially for that purpose. But you want eggs—the more infertile the better, and the more eggs the better.

I feed mostly prepared feeds of the best quality, believing that expert feed men should be able, with the aid of their scientists, to mix better feeds for particular purposes than I can. So I feed scratch feed in the litter in the morning with the object of keeping the hens working not only for warmth but that they may not forget their duty of the morning—the laying of eggs. At noon, a wet mash of prepared laying mash is fed with 5% of beef scrap and 25% steamed alfalfa and table scraps added. Mangel beets or cabbage are given as extra green stuff. At night another feeding of scratch food is given so they will go to roost with crops full of substantial grain.

It is rather difficult to say just how much scratch feed to give, but about 18 pounds seems to satisfy 100 Leghorns. Of the wet mash, 5 quarts to the 100 hens is enough. But should the hens not eat the food up clean cut the amount until they do not leave any. The noon feed of the laying mash can be changed to one of your own mixture if preferred. It is usually mixed about as follows: 2 pounds wheat bran to one each of corn meal, linseed meal, middlings, beef scraps; or substitute for the scratch feed 1 quart each oats, wheat and buckwheat and 2 quarts cracked corn.

The hopper should always contain grit of hen size, charcoal, beef scraps, and a dry mash of either of the noon feed ingredients, and oyster shells. Hens confined indoors must be furnished grit and oyster shells, which they cannot otherwise find as they could if at liberty, although oyster shells should be supplied even to hens outside.

Clean, fresh water is very essential. Particular care should be used to see that the hens have all they want to drink. Laying hens especially need plenty, eggs being composed of two-thirds water. Give your hens and pullets special care during the winter and feed them more than you would in the spring. This will seem hard sometimes, when eggs are so few it is hardly worth gathering them, but you cannot by any possibility get winter eggs unless your birds have all they need to sustain them and also produce the eggs. Farmers are very lax on this question. Evidently they feel they wont get eggs anyway, so the less feed bought the more they are in pocket. But they are wrong; for even if they did not get the eggs, their hens would be better nourished and so better prepared for early spring laying.

From October to February eggs are worth say 3½ cents average. It requires only that a hen should lay three eggs a month to pay for a good substantial ration. Five cents a month might keep the hen alive but an additional five cents worth of feed would probably result in her laying at least enough to pay her way instead of being carried at a loss.

CHAPTER XI

FATTENING BROILERS AND OLD HENS

The cockerels you intend to sell as broilers should be separated from the pullets as soon as you can pick them out and penned up in rather close quarters to prevent their exercising too much. Then they should be fed more with the idea of putting on weight as fat, than as bone and muscle; but my experience is that it is a hard proposition to make growing roosters very plump. The experts say they do it but the novice will have quite a job to add weight to his little cockerels faster than they would naturally put it on if fed all they need to eat. Of course, reducing the size of the exercising quarters and a special feed will help, and experience will also teach you how it may be done.

Last year I sold about 500 broilers of various breeds and found that the White Leghorn cockerels reached the 2 or 2½ pound size fully as soon as the heavier breeds, so up to this age the breed is just as profitable from the meat standpoint as any other breed. It is a question if there is any money raising beyond that weight when chickens would be sold in the fowl classification, because the price a pound is less. At 25 cents a pound live weight a good profit is made on the cockerels.

If you wish to try to add extra weight, the cockerels should be put into a house by themselves and wired up in a small yard away from the others, from two to three weeks previous to the time you expect to market them. The more quietly they are kept, and the less exercise they take the more favorable is the chance that they will put on weight. Then they should be fed all they will eat of fattening foods. I have used a prepared fattening mash for reasons aforesaid and have concluded that it will pay the beginner to get manufacturers' booklets on poultry and feeding. These contain full information as to the proper method of feeding chickens to obtain the best results for particular purposes, such as broiler market, squab market and to make heavy layers. Of course, these methods are all ways that call for using the prepared feeds; but if you prefer to mix your own there are various formulas.

HOME MIXED GRAIN FEEDS

Below is one, but the main idea is to give the fowls some fattening food they will relish and not tire of. Birds when reduced in their exercise are likely to lose their appetites. When that time comes they should be sold, as just then they will not add weight any faster than if allowed free range. Six parts of corn meal, two each of middlings and beef scrap and half a part linseed meal, mixed with milk or warm water and fed all the fowls will eat three times a day. The troughs must be removed as soon as the fowls are through feeding. As previously stated, I cannot seem to get much decided increase in weight, perhaps because my chickens are fed all they want when growing, and soon seem to go off their feed when placed in confinement on a mash feed.

Broiler raising and fattening is really a special branch of the business, so the beginner cannot hope to obtain even a good knowledge of all the lines of the poultry business in one year. He is again advised not to attempt too many things that might interfere with the more important part—the development and care of his pullets and hens. The time and study spent on the broilers results in getting only once, when they are sold, a few cents more on a chicken; but the pullets and hens are steady revenue producers and must not be neglected.

CHAPTER XII

WATER

All work connected with the keeping of chickens should be simplified as much as possible.

Water entails considerable labor as all poultry, especially laying hens, drink a great deal. It must be continually before them to keep them in good condition. It should be cool in summer and warm in winter and always fresh.

It is surprising how much water a large flock of hens will consume in hot weather—enough to make quite an item of the day's work—so any arrangement that will make the labor less deserves consideration.

A running stream on the farm is Nature's solution. Humans may have a barrel arranged on wheels so it can be pushed around to the various receptacles.

The photo shows my method of supplying water to the chicken houses.

This method may be modified or enlarged to suit individual conditions. The big barrel of about 200 gallons capacity, is set up on stilts so the bottom is higher than the troughs to be supplied. The water is pumped into the barrel by a gasolene engine, primarily used however for the water supply of my residence. It is then piped to various places as needed, the height of the barrel giving the

My method of supplying water to chickens.

necessary pressure to keep a continual small stream running.

The pipes should be buried a little in the ground as the water will become heated during the hot months and of course, this system must be cut out as soon as severe weather is expected and the pipes disconnected and drained out to prevent breaking by frost.

CHAPTER XIII

MARKETING THE EGGS

At least half of the success of a poultry business depends upon the profitable marketing of the eggs and fowls. It is generally conceded that most farmers spend much of their time and energy growing their crops, but not enough in finding a good, profitable market for their harvest. It is easy enough for any producer of farm stuff to sell all he can raise, but to sell his crops or eggs at a good price requires almost as much study and business ability as the producing of the stuff. A little difference of 3 cents a dozen on the year's output of eggs of the little plant outlined in this book would mean a clear difference in profit of about $40. Six cents a dozen would make an added profit of $80, which would make quite a decided difference in the profit a hen a year.

The middleman, so far as I can see, is a necessary adjunct to business, as a whole, but not necessarily so to yours or mine, so we will consider the wholesale market as the first one that would enter your mind as the place to sell eggs.

At the time these lines are being written, November, "near-by hennery whites," the best grade of eggs quoted in the New York market are listed at 60 to 63 cents a dozen. To the uninitiated this would appear to be as good a market as anyone would desire. So it would be if you had eggs to ship, but do you suppose for a minute that eggs would be 5 cents each at wholesale if there were any (comparatively speaking) to be had?

So far as the majority of small poultrymen is concerned (and large too) or otherwise the price would not stay up, the price might as well be $1 a dozen, for raisers simply don't get enough to make a full crate shipment in a month. To this extent this exorbitant price offered by the wholesalers is fictitious, making a good excuse to bolster up the storage egg price. On the other hand, when eggs are at the height of production, when everybody has the goods, and the wholesalers want them and know they can get them, they offer for the same quality 20 to 22 cents. Then farmers and poultrymen have the eggs and want to sell, in fact must sell, so these low prices are offered. After expenses are paid and returns made, the producer receives the magnificent sum of about 18 cents a dozen for the best grade of eggs in the market.

Let the experts who are delving into high prices with the object of finding who is responsible for the high price of eggs, or the women who buy, stop to consider that the eggs that go into market during the flush season by thousands of crates, go in at a price which would afford the producer not much more than cost if he had his hens enclosed and was compelled to buy their entire feed; and there should not be much difficulty in placing whatever blame for high prices there may be, where it belongs! But an unvarying price the entire winter, appears to me the opposite extreme. It also should be a matter for expert investigation, rather than for a body of women with practically no knowledge of the egg business to get together and declare the arbitrary price of 30 cents a dozen. You can get a good price anywhere for eggs during September to February, but the more important thing is to get a good price from February to September, when you and everyone else has eggs to sell. Here is where you need to get some of that "business ability" at work. So, the first market, the wholesalers, may as well be passed by if you expect your books to look encouraging to you at the end of the year.

There is the hotel trade. Perhaps most novices have this market in their minds as the road leading to great profits, believing that as soon as their plants are in operation, all the hotel buyers will come begging them for eggs at any old price they will ask. Do not base your hopes too high on the hotel trade for "oft expectation fails, and there where most it promises." The hotel buyer is like most other people in that he wants all he can get for his money, and he is in a better position to get what he wants than the average user of eggs; for, being a large buyer of not only eggs but other articles of the food, he can get better prices, and better service than can a small consumer. So the small poultryman will find that if he gets any offer at all it will be so mighty close to the wholesale price that he would hardly distinguish the difference. Besides a steady supply would have to be guaranteed that would bother the beginner who has not had at least a year to study conditions. An exception might be found, however, in some hotel willing to pay a fair price. If such a buyer were found, that market would be at least fair, as one fair sized hotel would use the entire output of a small poultry plant, thus saving some money in shipping and bookkeeping details.

THE PRIVATE TRADE

As a general proposition there is one and one only, really good, profitable market for the eggs produced on a small farm; namely the private trade. The private customer, using individually, small quantities of eggs, is at the mercy of the retailers, there being no inside price for her. If she wants the goods she must pay the price, taking chances as to quality. If the beginner decides to go after this trade, the thing of primary importance is that he shall, at all times, keep the quality of his eggs up to the highest standard. No matter at what price he sells, the eggs must always be just what he claims them to be. This is the only way to build up a profitable family trade. So great care is needed to avoid any mistakes that might lead to some bad eggs, as one bad egg might lose a customer, who if kept, would use dozens of eggs a year. "A bird in the hand is worth two in the bush." In the egg business one pleased customer will always lead to others, and acting as a link in an endless chain will probably result in new customers keeping up to increased production.

Two years ago I took four dozen eggs to my former home and sold them to the neighbors. From that time I have not had to advertise or solicit orders, as my customers did the advertising for me. Since then I have shipped home 50,000 eggs. I have not asked as high prices as some say can be had, believing that a ready sale with no surplus left on hand to worry about is better than holding the eggs for higher prices.

The consumer who honors you with his or her custom, should certainly have some share in the saving due to the method of doing business. The great argument of many for the lowering of prices is the elimination of the middleman. While that seems to be an impossibility except to a limited extent, still, if you sell direct, consumers are entitled to some division of the profits that would otherwise go to a couple of middlemen; otherwise their advantage in doing business with you is only the fact that they get a good article. The middleman in some form, however, in general business, cannot be done away with for the needs of some families are so small that it would be impossible for small shipments to be made profitably either to buyer or seller.

Again, this is made impossible by the fact that in the spring there is a great surplus of eggs beyond the daily consumption. Unless such surplus is absorbed by the storage house great loss would result as eggs deteriorate very rapidly in warm weather. The poultryman wants no accumulation on hand if he is catering to the family trade.

Part of the storage eggs may come from poultrymen who send in their surplus rather than hold for a length of time, but the vast majority probably are gathered eggs. All through the country are stores and dealers who collect eggs, the former taking eggs in trade for groceries and other merchandise, allowing 15 to 21 cents a dozen in the spring. These eggs get to the store house some way. A fresh laid egg put in cold storage the day it is laid would probably be just as good six months from that time. But how many are put in storage in the fresh condition? Farmers save up their eggs until 100 or 200 are on hand, then sell to the store or the collector. When the dealers' stock has reached several hundred they are crated and sent to market, the majority being far from fresh when they arrive.

These eggs which bring 18 to 20 cents a dozen, are candled to determine their condition. Some are thrown out, some broken, and the others graded as to their condition from first class down to mighty poor class. They are then put in cold storage, and four, six or eight months later, the consumer buys them in the hope that they are good. Cold storage, however, is indispensable. While there are some bad features, let us give the business its due credit, for without it the public for six months would have to do without any eggs or pay exorbitant prices.

From September to Janaury is the time when fewest eggs are laid. This accounts for the high prices. Hens are molting then, old hens laying practically none at all, and most spring pullets have not begun laying, hence there is a practical halt in production. Some writers, (one in particular) claim that their pullets are brought to laying age just at the time the older hens stop, so they go right on filling the orders about the same as usual. I will not dispute this assertion; but while we have all had some pullets that lay early, still, to have a whole flock take the place of the molting hens seems hard to believe. As the personal statements in this book are facts, so I believe these other statements are honest. Although pullets' eggs as a rule are much smaller than mature hens' eggs and I wouldn't care to send all pullets' eggs to my customers, still I put this statement before you. If you as a beginner can accomplish what they claim, you can add about $1 to the average profit a hen as claimed here.

METHODS OF SHIPPING EGGS

To return to your private trade customers. There are several ways of shipping eggs to the family trade—parcel post, express in small lots and express in larger lots. Parcel post is hardly satisfactory for an economical method of distributing eggs. Small lots of six dozen or so to each family is practically the same method as the larger lot way, the differences being in the saving of boxing and detail to the shipper when 30 dozen are sent, as against the sending of five dozen cases, and the saving of expressage to the consumer in the same proportion.

Six dozen eggs cost 27 cents to express 100 miles, while 30 dozen cost only 42, so the express cost a dozen runs from about 5 cents

MARKETING THE EGGS

down to a little over one. If the shipper could induce several families to take a crate of 30 dozen eggs to be divided up upon arrival, it would pay him to prepay the charges. Thus the mutual saving of expenses would result in the cheapest price at which his eggs could reach the consumer, and he would be much better off by the saving which would result from packing, labeling and delivering one case, instead of three or five.

The packing of these eggs, whether in 6, 12, 20 or 30 dozen crates is a matter of great importance. The wholesale trade requires eggs to be packed in regulation 30 dozen crates, with cardboard fillers and partitions. Why half the eggs are not broken in transit is a mystery. That many are broken must be the case as the government is investigating the great losses with the idea of offering suggestions looking toward saving the waste, as any loss by breakage is a direct addition to the cost of living, because the breakage has to be reckoned into the ultimate price to the consumer.

Any method that will save this breakage is of value to the shipper. As I ship only to private trade I have my way, which although entailing more labor in packing, still pays for the trouble many times over, as out of 50,000 eggs shipped during the past two years, but one customer has made a claim against the express company for breakage, and that claim for only three dozen. This method is worth many times the price of this book to you if you are to ship to the family trade, as that trade above all others desires to avoid a nasty mess of broken eggs.

CHAPTER XIV

HOME MADE CASES

I make my own cases of 6, 12, 15 and 30 dozen size, of ⅜ inch box boards, the same dimensions inside for the 30 dozen size as the regulation crate and the other capacities in proportion, except that they are about 1 inch deeper. On these crates are stenciled my name and address making a neat advertising crate. The difference comes in the packing which begins with about 1-2 or 3-4 inch of sawdust with the card board division set on top. Then the fillers are put in and the eggs placed in until one layer is complete, after which all the spaces between the eggs and the sides of the crate are filled with sawdust, packing it in with the fingers until tight. Other layers are added and treated the same way until the top is reached, when the remaining space between the eggs and the top is covered with sawdust. A card board division finishes the job.

This makes a solid mass of eggs and sawdust, with no motion of eggs from side to side or up and down. One can be reasonably certain that those eggs will reach the customer as sound as when they were packed. Only dry, soft wood sawdust should be used. The crates weigh more than the regulation crate, but are delivered by the express companies for the same price, 30 dozen being delivered in New York for the 42 cents. The express company has been the gainer because there has been only one loss by breakage.

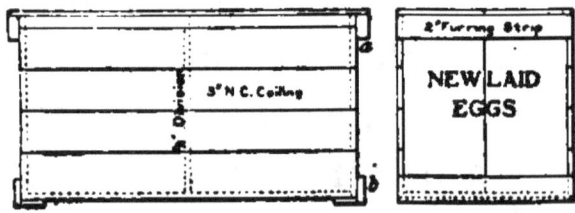

Details of home-made egg crates.

CRATES AND SHIPPING

To ship in this way you should buy 1,000 fillers, 3 dozen size and 1,000 card board divisions; the fillers costing about $7.00 per M and the divisions $2.50 per M. By having the crates returned these fillers and divisions would last the ordinary poultryman a couple of years at least.

Dimensions for Egg Boxes and cost of shipping, etc.

Of the crates in the following table the 6, 12, 15 and 30 doz. are the only ones used by me.

No. of Doz.	Length	Width	Height	Fillers	Layers	Box Cost	Divisions & Fillers Cost	P. P.	Ex.	Cost Per Doz.	
										P. P. 7½	Ex. .15
2	12	8	4	1-2 doz.	1	4	1	10	25		
4	12	8	6	2-2 "	2	5	2	15	25	5½	.08
6	12	8	9	3-2 "	3	6	3	21	25	5	.06
8	12	8	11	4-2 "	4	7	4	25	28	4½	.05
10	12	10	11	4-2½"	4	8	4	31	29	4 10	
12	12	12	12	4-3 "	4	15	4		35		.04
15	12	12	13"	5-3 "	5	27	5		35		3.2
20	16	12	12	4-4 "	4	36	6		35		2.5
24	25	12	13"	8-3 "	4	42	8		40		2.25
30	25	12	13"	10-3 "	5	53	10		42		.02

In the above table three dozen fillers are used cutting off part where necessary: i. e., the two dozen size would be a three dozen filler with two rows cut off, the 2 1-2 dozen would take a three dozen filler with one row cut off and the four dozen would be a three dozen filler with a one dozen set in alongside.

It will pay to have any crate above the four dozen size returned as the express companies charge only ten cents for empty crates

My method of packing eggs for the private trade.

even up to the 30 dozen size. Any crate made in a substantial way should be usable at least ten times.

Figuring on this basis the actual average cost of delivering the 30 dozen eggs would be one-tenth the cost of crate and fillers 6.3 cents, expressage 42 cents, return of empty crate 10 cents, a total of about 60 cents or 2 cents a dozen.

All eggs should be examined before shipping, dirty eggs washed, cracked eggs removed and under sized eggs saved for your own use.

If at any time you should deliver eggs in small lots of a dozen or so, they should be packed in boxes holding one dozen each, these boxes having your name and address on the cover, and making a neat way of advertising your business. If you ship direct to the families in the way I do, you cut off this expense of the boxes which means almost one cent a dozen. Pretty boxes are all very well, but the main things are to maintain the high standard of quality and to have the eggs arrive in perfect condition. We have all bought eggs in nice artistic boxes labelled "strictly fresh eggs," etc., and later have found that the box was the best part of the purchase. All the egg crates I have made and shipped are returned empty by the express company for ten cents each. They can be reshipped many times, reducing the average cost a crate to a very small item.

CHAPTER XV.

STORING SURPLUS EGGS

One of the greatest problems of the egg producer is to get a market in the spring for ALL his eggs at a good price. It is probably the same problem in another branch that the farmer has to contend with in all his productions. In the egg business I believe there is only one way it can be solved and do justice to farmer and egg producer. That is by establishing small local storage plants on a co-operative basis by which system the farmer could take his eggs to storage during the flush season of spring, get his cash at the market price and then when his eggs are sold later from storage get some extra or pro rata share of the profits which now go into the hands of speculators.

Most farmers and egg producers need the cash in the spring to pay their bills and so must sell their eggs as they can whatever the prevailing price may be. It is a pity that the producer who has his money invested in hens and equipment and who devotes all the year to their attention must sell at 18 to 20 cents during the flush time to speculators who hold the eggs for five or six months in cold storage at a slight cost outside their money invested, and who then sell at 28 to 30 cents.

One spring my production ran beyond my market at fair prices so I located a storage house near-by and sent my eggs which were only one to three days old, into storage to be held for my own account. I accumulated 63 crates which were held in storage 2 months on the average and sold them at 30 cents a dozen, instead of the 18 or 20 cents which was the prevailing price during April and May. The cost a dozen including new crates, expressage and storage was 3½ cents a dozen. You see where the dealer who keeps and places eggs in a storage house makes his profits.

Putting eggs into cold storage, should be however, the last resort of the egg farmer because it would necessitate tying up capital for which there is plenty of need in the spring.

The safest and most profitable way is to keep figuring ahead, endeavoring to have your market keep pace with the increased spring production even if the eggs had to be sold at a smaller margin of profit.

Should you be compelled to put a surplus in storage, find a modern dry air plant in preference to an ice cooler as the proper results can only be obtained with a dry air. Ice coolers are liable to create a damp air which would tend to make your eggs musty or moldy.

All eggs that have been stored for any considerable length of time should be candled before being shipped to the consumer as every egg even in one day's collection may not be alike in condition.

During the busy laying season several hens may select the same nest in which to lay and no matter how many nests you may have prepared there will be some selected as favorites resulting in 6 or 8 eggs being laid in one nest.

This would mean that the first egg laid would be set on several hours by the layers of the later eggs thus making a slight difference in condition of each of those eggs.

When these have laid in storage for six months that slight difference might be accentuated and a change take place that would require those eggs to be candled.

The above applies especially to fertile eggs which are very quickly affected by heat and require great care in their proper handling.

This storing may be accomplished in several ways, all of which are more or less good according to the length of time you desire to keep them. Eggs may be kept a month or so if they are packed in boxes in oats, bran, buckwheat chaff or ordinary salt, and the boxes turned every few days. The eggs stored by any of these methods will lose weight by evaporation. If packed in salt this evaporation is likely to cause the salt to cake, thus making it difficult to remove the eggs. The storing by any of these methods requires that the eggs be kept in a cool, dry place as the packing would absorb dampness and so taint them.

The most satisfactory and by far the easiest way to preserve eggs for family use is by the waterglass method which is as follows:

In a clean five-gallon crock, pack clean fresh eggs either filling the jar at one time or packing a few each day. Boil 10 quarts of water (preferably rain water). After it has cooled stir in 1 1-2 pints of commercial waterglass (sold by druggist). Pour in this solution until it completely covers the eggs to the depth of say half an inch. Now the jar may be set aside, with a board cover to keep the dust out, and no attention need be given beyond adding a little water from time to time to replace the loss by evaporation. Eggs may be used as wanted from the top. In December my family used eggs that were packed during April and May. It is essential for good results, that all eggs be fresh, and better if also infertile.

CHAPTER XVI.

TESTS FOR FRESH AND STALE EGGS

Here are some tests that may be made to distinguish a fresh egg from a stale one, provided the eggs have not been in preservatives:

Eggs that have been in waterglass months will stay at the bottom when immersed in a pail of water, but eggs 10 to 14 days old that have been kept in a room with a temperature of 65 degrees will turn up on end and begin to rise to the top of the water as the age of the egg increases, so a stale egg will come to the top. If kept in a cool place it will be nearly a month before any such change occurs.

Again if an egg kept as above is candled or held to the light an air space will be seen after a couple of days. The air space increases in size as the egg ages. Candling is the surest method of determining the quality of an egg as every defect an egg may have is shown by candling.

Another test is to shake the egg. If stale the contents will run together so one can hear the motion in the egg. This test does not apply in winter months as a new laid egg may freeze and upon thawing out it will shake.

Not all stale eggs, however, are bad. For an experiment, I took an infertile egg from my incubator after 10 days incubation, had it hard boiled. If the circumstances connected with it had not been known, I doubt if anyone would have called it bad, for it really tasted good.

A new laid egg eaten the day of laying or perhaps the next day, after boiling, will show that it is new laid by the skin adhering to the shell: it will not peel easily.

CHAPTER XVII.

MARKETING OF BROILERS

To the poultryman who makes a specialty of table eggs, the sale of broilers is simply a side line; yet it is a part of the business that adds considerably to the year's income. Half your young chickens will probably be roosters. Since these little fellows can be sold as broilers when two to three months old, it is the first revenue that you will receive from your spring chickens. In addition to your own stock it would be very easy to buy up little roosters from other smaller producers, so the broiler end of the business could be made quite a large item.

Broilers are sold alive as well as dressed, some preferring one way and others purchasing only when dressed. For the poultryman catering to the egg market, the market taking the broilers alive, would seem to be the most satisfactory since it requires little time to prepare the birds for delivery. At the season when time is an important factor this is well worth considering. I sell my broilers alive at 25 cents a pound. At the weight of 2 1-2 pounds this price is about equal to 30 to 31 cents a pound if dressed, and entails no work beyond crating the birds. If, however, your market prefers the cockerels dressed, you will have to learn dry picking, as butchers and hotels prefer, in fact almost insist on dry picked chickens. As there is quite a knack in picking this way, you would have to learn the art before contracting to deliver that way.

It has been mentioned several times in this book that your first year is understood to be a year of study and preparation for the following years. In the selling of eggs and broilers you will find this year's experience will be necessary to you, to learn just what you can count on in order that you can agree to fill contracts that might call for specified quantity at regular intervals of time. An order for 30 dozen eggs, or 50 broilers a week, would be easy during the productive season, but before you agree to keep sending these products regularly, you will have to learn the limitations of the business and your ability.

FALLACIAL RECKONINGS

The first poultry book I read setting forth the fortune to be made from poultry had it all figured out that each week the plant would ship 40 or 50 broilers as regularly as clock work. The incubators would always hatch the same way, the brooders and everything always working in unison toward success, always good luck, with regular income each week. There was no chance of failure or disappointment! These statements may be useful to all of us encouraging us to bigger things; but after you have put in your first year, your books before you, see if it would have been safe for you to have started business, using some of these figures for your basis of figuring!

Be assured, however, that you will be able to sell your broilers even the first year. Sell them just as soon as they have arrived at the proper weight, as it will not only return you some money but every cockerel sold is one less to feed and look after and perhaps interfere somewhat with your care of the pullets. The pullets are to be your future source of revenue, and should have every care and attention that will help to develop them into strong early laying birds.

FEED COST OF EGGS

According to my books for many months, the cost of feeding, if all the feed has to be bought, is 9 to 10 cents a month a hen. I feel safe in figuring at this rate. It is the basis of the estimate made for keeping the 200 hens as in the plan laid down in this book. I have had to cart all my feed seven miles from the station at a cost of 10 cents a bag.

CHAPTER XVIII.

SELLING OFF OLD HENS

Each flock of pullets should be banded with leg bands when put into winter quarters, so the age of any bird can be told by looking at the banded leg, in order that the following summer, when you look over the flock with the idea of selling the surplus hens you will know the age of every one. Until your flock has increased to the size that satisfies you it will not be advisable to sell yearling hens. Anything older should be disposed of unless some special ones are to be saved for breeders.

If you have some strain or breed that for some reason—that they are great layers, for instance—are well-known, you may be able to sell the hens for an advance on the price they would bring for meat; but unless you are fortunately located some place where your neighbors would purchase these hens, it would probably require some advertising to enable you to sell beyond regular meat price, but if your $1 hens have paid you $1.50 to $2.00 the past year, you can afford to sell them for a little sacrifice from their cost price. Hens should be sold at two years of age because if properly fed and handled they have seen their best days at that age.

It is stated as a scientific fact that when a pullet reaches maturity she has in embryo all the eggs that she will ever lay; that if forced for laying, the first or pullet year will be the most prolific year; and that after two years her laying will be of a desultory character. If a pullet is brought to laying age in October there probably would be no doubt that she would lay more eggs than a yearling hen; but whether the pullet would lay more large, saleable eggs than the other is a question.

Pullets, when they first commence laying, as a rule, lay small eggs, eggs of a size that might hurt your conscience to send to your family trade at high prices. A pullet lays smaller eggs than mature hens do because she may commence laying before she is fully developed as to size and weight and because some more of her food is required to keep building up her body thus leaving less for the production of eggs as would be the case with the yearling hen.

SELLING OFF OLD HENS

Again, nature has provided that the eggs have to leave her body between her pelvic bones. In a pullet these bones are so close together at the time she begins laying that two fingers will scarcely go between them, but as she continues to lay they are gradually forced apart so that in a yearling hen sometimes three fingers can easily be placed between them. In this case a hen is said to be a good layer of large eggs.

The price at which fowls are sold does not vary much unless some special market is found. No poultryman cares to sell fowls in the spring so I imagine the chickens in the market at that time must be cold storage. Directly before the molting season is the best time to sell as the hens have then finished their profitable season. If carried through the molt the income will result in just that much loss. At this time, July and August, the summer boarding season is on, so a location in proximity to a summer colony then shows its advantages as fowls bring 18 to 20 cents live weight at just the time they should be sold.

CHAPTER XIX

WIRE RUNS AND DELIVERY CRATES

It is always handy and sometimes necessary to have a couple of wire runs to use, to keep chickens to one spot, either for protection from hawks or cats, or to get them accustomed to a certain house. Make the ends of 3 inch stuff to strengthen the whole frame, as it is cheaper to make well rather than have to keep repairs. The sides and braces can be made of 2 inch furring strips, the braces nailed at the proper distance along the sides to take the width wire you will use.

If made 6 feet long, the brace will be in the middle so 3 foot wire goes around just twice. If 8 feet then the braces should be 4 feet from the ends to use 4 foot wire. The 9 foot runs should have 2 braces spaced 3 feet apart. The end, which is to set up against the house, should be left open as well as the bottom, but the other end should be wired tight. It is more convenient and makes it less likely that the chickens run out, if this end is permanently closed and the door, through which the chicks are fed is put in the top. These runs can be of any width, height and length. They will last for years if made right at first and painted.

I made a delivery crate which just fits my wagon and will hold 100 live broilers. A

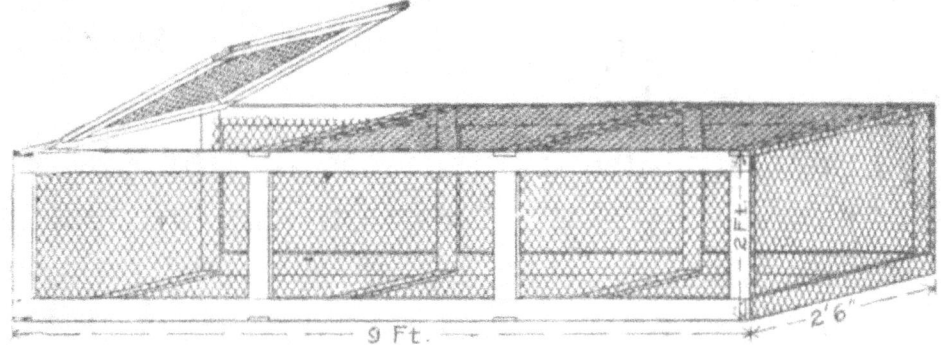

Wire runs, well made, are more economical than those cheaply constructed.

partition divides them so they cannot overcrowd. Doors 8 inches square are put in the top so the chickens can be reached but they cannot fly out while an arm is reaching them. A seat is fitted on front. Do not crowd broilers or fowls in any delivery or shipping crate, especially in hot weather, else some will be sure to smother.

CHAPTER XX.

FEED COST OF EGGS

Last year my flock laid 36,000 eggs, the production being distributed during the months as follows: March to July, 74%; August, January, February, 16%; September to December, 10%.

This distribution of the production plainly shows why eggs are cheap during the spring months and high at other seasons. Some say the only way to make money from chick-

Combination shipping crate wagon seat for delivery broilers and fowls.

ens is to get eggs during fall and winter. While this is one way and a hard way too as pullets don't always lay as one would hope,

FEED COST OF EGGS

still there is another way; namely, by good salesmanship and good eggs, to keep the spring price up to what good eggs are worth.

Eggs do not spoil quickly in cold weather but deteriorate rapidly in warm weather. Customers who realize this will pay a premium for new laid eggs in the spring. Now if you cannot either get a good production in winter or a good price in the spring, make up your mind that poultry will not pay.

New laid eggs to city customers should bring not less than 30 cents even in April and May. My flock of 375 averaged about 100 eggs a hen. The feed cost a dozen eggs was about as follows:

	Cents
3000 dozen cost 375x$1.20 a hen...$450 average for year	15
10% or 300 dozen in 4 months	50
16% or 480 dozen in 3 months	23
74% or 2200 dozen in 5 months	8

Now if I can increase my average production to 150 eggs a hen it will reduce these cost prices one third. These then are the two objects to work for: an increased average production and fair prices for the eggs at all times.

CHAPTER XXI

CHARACTERISTICS OF BREEDS

Hens laying white eggs are the Leghorn, Black Spanish, Campine and Minorca, the last laying very large ones. The layers of brown eggs are Brahma, Cochin, Wyndotte, Langshan, Orpington, Plymouth Rock, Rhode Island Red and Dominique. Langshan and Minorca, about 30 ounces to a dozen. Leghorn average eggs about 24 ounces to a dozen. Pullets average eggs 16 to 20 ounces to a dozen. Bantam eggs weigh about 14 ounces to a dozen.

Eggs are composed of 11% shell, 32% yolk and 57% white. Contrary to ignorant popular belief the food value of eggs is not affected by the color of the shell nor the color of the yolk. The shell color is peculiar to the hen laying the egg but the color of the yolk is dependent greatly upon the kind of food. While the color of itself does not effect the flavor the food does and eggs may be made to taste strongly of some particular food the hens eat, such as onions.

Eggs to be marketed fairly, should be sold by weight as they vary greatly in size, some running as high as 31½ ounces to the dozen. Others are offered for sale weighing 20 and 22 ounces. The food value would not vary as much as this for the reason that the yolk in medium sized eggs is nearly the same size as the yolk in larger eggs, the extra size being made up of white which is four fifths water, while the yolk is rich in fat, iron and lime salts.

CHAPTER XXII

BREEDS AND WEIGHTS OF FOWLS

The various breeds of chickens with the standard weights of each are given below:

STANDARD WEIGHTS OF FOWLS

Breed or variety	Rooster	Hen	Cockerel	Pullet
White / Brown } Leghorn	*No standard weight but runs about* 5	4	4	3
White, Buff, Silver Laced	8½	6½	7½	5½
Orpington	10	8	8½	7
Rhode Island Red	8½	6½	7½	5
Black Minorca	9	7½	7½	6½
Light Brahma	12	9½	10	8
Cochin	11	8½	9	7
Plymouth Rock	9½	7½	8	6½
Langshan	10	7	8	6

CHAPTER XXIII

GROWING OWN FEED

I have had many persons say they should think there would be much more money in raising chickens if the poultryman raised his own feed; but as a business proposition I do not see it, at least only to a small extent. A flock of 500 chickens could be kept on an acre or so but to raise the feed for such a flock would require many acres. Even then the entire variety of feeds could not be grown. Raising chickens in quantities is a specialty as is farming or the raising of the grain.

A large flock of chickens requires all the attention of one man or more according to the size of the flock. Any farm work done would require just that much extra help.

Every acre planted to corn or wheat would be an investment for just that work and if the additional cost of the farm acres together with the additional help, barn, horses and tools were charged to the cost of the feed it would be the plain proposition that the saving would be just the difference between the cost of producing by the farmer and the cost in the open market. From the way the farmer talks of his hard lot, the difference cannot be considerable. To my mind it is not sufficient to pay for the division of energies and attention by which the poultry might suffer.

We will agree that should a man have a large farm with fertile acres and a knowledge of farming, the necessary capital, etc., he might save some money by growing his own feed in the same way that a manufacturer of clothes might make a little more money if he were also the manufacturer of the cloth that goes into the clothes; but this double business is no more necessary to the one than to the other, and few men are able to make a success of more than one business.

CHAPTER XXIV

COMMON DISEASES OF CHICKENS

Watch should be kept in October and November for roup, one of the deadliest diseases of poultry. Roup is caused by drafts, cold or a damp house. The first evidence shown is running at nostrils and eyes. As the disease develops, cheesy matter forms in the eyes and at the roof of the mouth, one eye begins to swell and close up, and if the disease is not arrested, both eyes are closed and the hen is done for.

At the first sign, remove the sick bird to some outside house with good, fresh air and doctor her, for the disease is very contagious, spreading through the drinking water. There are various roup cures, application of which are made directly on the affected parts, and also by stirring in the drinking water as a preventive. Do not neglect immediate treatment if you expect a cure, and do not breed from any fowls that have been affected. I have used commercial roup cures more as a preventive than as a remedy, as I have found it almost impossible to save a hen once the disease has made much headway. "An ounce of prevention is worth a pound of cure," so above all things keep your house dry and clean and begin adding the roup cure to the drinking water before the disease shows itself.

Lice and mites cause much trouble as well as loss of chickens. Lice do not really endanger the lives of the hens but are a great annoyance, as you will notice if you watch a flock of hens pick the insects off with their bills. Lice stay on the hens, living on the scale of the skin, etc., but not sucking any blood from the chickens. If the hens are supplied with a dust bath of sand or ashes after they are enclosed for the winter they will not be seriously troubled. They will find some place to dust themselves in spring and summer, so lice will not affect them.

Mites are a more serious menace to the health and vitality of fowls, as they work during the night when the hens are defenseless upon the roosts, sucking themselves full of blood and then sneaking off in the morning and hiding in any crack or crevice that is handy. I have seen them in old houses, hiding behind some box on the side wall or under some roost so thick it looked like a spot of red paint. When they were struck with a hammer there would be a spot of blood. Hens cannot be expected to do their duty under these conditions. The result is reduced egg production and even death of the hens.

All coops housing old fowls should be sprayed with kerosene, using a hand spray pump, which costs about $3.75. This powerful spray will get into all crevices and if continued regularly will at least keep the mites down to a point where they will not be dangerous.

Farmers are troubled more with lice and mites than regular poultrymen are, because their coops are not as accessible to cleaning and because the farmers are often neglectful. Setting hens are bothered terribly with these pests, often being killed right on the nest, but more often leaving the nest with a hatch half over.

Scaly leg is also caused by a similar mite parasite, which burrows beneath the foot and leg scales and gradually forms a crust or scale which is very unsightly looking. Since scaly leg readily spreads from fowl to fowl never put a hen having this trouble in with the others.

If the house and the roosts are regularly sprayed and disinfected scaly leg will not appear. Such hens as already have it may be cured by swabbing on kerosene, making applications until the scales are removed.

Bumble foot, a swelling which forms on the under sides of the hen's foot, is caused by the fowls jumping down upon a hard floor, which has insufficient litter on it. The leg bone of the hen seems to end at just the point where the foot meets the floor without any provision being made to absorb a jar. This bone strikes the hard floor. If the bird continues jumping down from a high roost, a boil or swelling begins to form between the toes, often growing to a large size. The prevention is easy—simply deep litter on the floor so there will be no jar. If the hen has not contracted a severe case the swelling will gradually disappear.

Diarrhea is caused by improper feeding or unsanitary conditions. Birds on range with natural conditions seldom have diarrhea, but yarded fowls which have to depend upon the attendant for all their food are often afflicted.

Lack of supplied grit is one cause. Too much meat often causes it and a ration of too much corn will also do it. Change the ration, clean the house and put 10 grains of sulphocarbolate of zinc to the pint of water. It should cure the birds. Diarrhea is evidenced by the color of the droppings, which become yellow and red in streaks as well as thin and watery.

Gapes, I have never had in my flock that I know of. It comes from the fowls eating worms from the soil of a chicken yard which has been used as such for some years. The worm establishes itself in the wind-pipe of the chicken, causing irritation that the bird tries to relieve by stretching its neck and coughing. Young chicks are mostly affected, old fowls seldom. The worm is removed with a horsehair, a feather or other small hook, but the removal often injures the chick. Numerous other diseases affect chickens but to distinguish them will require study and experience.

A pen should be provided as a hospital for any sick birds, which should be removed to it as soon as seen to be ailing; for one sick hen may cause a lot of damage.

CHAPTER XXV

ROUTINE WORK IN THE SPRING

It may be well to give a little idea of the daily work in the spring so a novice may figure as to whether he ought to grow his own feed or run a farm as a side line to take up his spare time. Say the plant is one of 500 fowls and 500 young chickens with incubators set for more.

Each of the items below—water, feed, houses and lamps would mean about 10 of each.

SCHEDULE OF DAILY WORK

6 A. M.—Open all houses.

6.30—Feed old fowls, water old fowls, feed young chicks, water young chicks.

8 A. M.—Turn eggs in incubators, fill lamps in incubators, fill lamps in hovers.

10 A. M.—Water all chickens.

12.30—Feed all chickens, collect eggs.

5 P. M.—Feed all chickens, water all chickens, collect eggs.

7 P. M.—Turn eggs in incubators, see all chickens get in houses, close and lock all houses.

In addition to the routine work to be done every day would be the following to be done as needed:

Keep all hoppers filled with feed, clean all houses, kerosene and disinfect as needed, pack eggs, some for market, others for waterglass, make egg crates, test eggs, deliver eggs and chickens to market, get feed, take care of hatchy hens, pen off breeding hens, build fences, take care of new hatches, etc. Before the day is over the poultryman will find he has had a long, hard day's work, ending with half an hour's exasperating labor of closing in the chickens for the night.

Spring is indeed a busy time and much of the work is very important, as on it depends the successful growth of the young chickens to be the next year's stock.

Every night I have before me a list of special things to be done the following day. On many days the list is unfinished at night. Most of this work, however, is in the open air. You have the pleasure of watching your flock growing and the hopes of success that all the work brings, so the time passes rapidly enough, even if you have put in a thirteen-hour day. Non-union hours prevail in the chicken business!

CHAPTER XXVI

GENERAL REMARKS

An additional source of revenue is the manure which if taken of and stored as already described will sell for considerable money in the spring when farmers sow their corn, etc. Little as well as large things help to success. The manure item is one of the small things that will help your books to look well at the end of the year.

Another source of profit, provided the land is your own, is the extra growth that fruit trees will make on ground run over and fertilized by the droppings. If you purchase a place that is not already blessed with fruit, plant the very first year trees in the hen yard so as to gain every year possible in growth and age at which they will bear fruit. I have 15 two year old fine, large, thrifty peach trees which average 8 inches circumference, 6 inches from the ground, and 25 set out this spring. Two of these latter measure 5 inches in circumference, 6 inches from the ground.

The others are not far behind, but all of them far beyond the others set outside the yard.

Hen manure is one of the best fertilizers. When to the droppings which enrich the soil, the scratching and cultivating of the hens is added, you have an asset added to your orchard that means considerable. The growing orchard in return, gives shade to the chickens during the hot summer weather.

On the other hand, do not plant berry vines or currant bushes or any other low stuff, or the hens will clean out all the foliage and leave nothing but bare stalks. Chickens are crazy for green food, especially during the summer when most plant life is either dried up or tough. No vegetation will last long when a flock of chickens has access to it.

The question may be asked as to how much space is required to keep 500 hens. The answer is that 500 hens can be kept on a quarter to half an acre, in a house 100 feet long with a yard of 100 feet, which would be one-quarter of an acre; but 500 little chickens could not be properly grown on less than an acre.

The past season I had 400 hens in a long laying house with a small yard. Then I raised 900 little chickens to maturity on two

GENERAL REMARKS

acres for the entire flock, but more space would have been preferable. Growing chickens need plenty of range to build up good strong constitutions to fulfil their mission in life; just as boys and girls need active outdoor life to develop them into robust men and women. Later in life these children by reason of their early, healthful lives, are able to stand confinement in offices and factories on the same plan that old hens can be confined to small space and still do well.

Hens to be used as breeders, however, should be housed under the same favorable conditions that made them hardy pullets, for they are to be the layers of the eggs from which other chickens are to be hatched, therefore, need to be rugged and healthy to the last degree. To keep their good health and not impair their constitutions, they must be kept under the best conditions—those nearest to Nature—life in the open on the ground.

Last winter I put 15 hens and a rooster in each of four pens 10x12 feet in one end of my laying house, for breeders, selecting of course the best. They out-layed, in comparison, the balance of the flock in the large room, proving the previous remarks under "laying house." For breeding purposes, however, the eggs were very poor. Later, when they were put outside on the ground, the eggs were better and the change had especial effect upon the roosters. The 6x6 foot colony houses would make ideal pens for breeding flocks.

HOW TO GET RID OF RATS

One of the worst enemies of the poultryman is the rat. Rats will do considerable

Raising a Colony Coop off the ground to prevent rats nesting under same.

damage among a flock of young chickens, and should be got rid of as soon as their presence is known. It is a difficult matter to catch a rat, especially when the rodent has his home among the chicken houses. Traps and poison that might otherwise be used,

GENERAL REMARKS

cannot be here, because you are as likely to get chickens as to capture rats. There are poisons or fluids that act the same as poison to rats but are said not to affect chickens. I used one of these last fall with good results, cleaning out a family of rats in my cellar. This exterminator causes the rat eating it to contract a disease peculiar to rats, the disease being passed along to all the other rats with which he comes in contact, resulting in their death. If all houses are raised off the ground as mentioned under "colony house," and all houses floored, and then properly closed at night, rats will not bother you. The only rats that bothered me were under houses built with raised floors, but enclosed all around to the ground.

GOOD METHOD OF FENCE SETTING

The best way to put up a 6 foot wire fence so it is practically free from "bags" is to set the posts, allowing them to project 6 feet 6 inches above ground. Then unroll the wire the full distance to be fenced, even to the 150 feet in the roll, letting the edge that will be the bottom be against the posts. Hang one end up on a nail at the top of the post, then continue on down the line of posts hanging the wire on all. When the length is hung up, tack one end at the top only, straightening out the top in one nice even line all the way down the line, then go on tacking to the posts. Now put the bottom board on the line with the lower edge of the netting, tacking the wire to it and you will have as nice smooth fence as can be made. If there are inequalities in the ground set the board in or fill the ground up. You can never make a good job by trying to run the wire up and down over the inequalities.

The above applies to the common 2 inch mesh chicken wire, but I have found that the strongest and best appearing fence is the kind made of straight, vertical and horizontal wires, the lower part spaced 1 inch, with the spaces widening up to the top.

One of the handiest appliances to have on the plant is a bundle or so of furring strips, ⅞x2 inches, planed three sides and costing ½ cent a foot. These can be used in so many places, such as for roosts, wire runs, etc., that they are almost indispensable.

Another very useful thing, making the capture of a hen a very easy matter, is a stick with a piece of telegraph wire attached to the end, and bent like a shepherd's crook, the

GENERAL REMARKS

Hen catching crook in use with details.

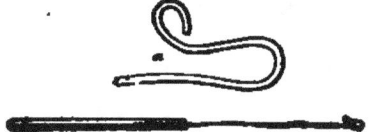

crook being about 3 inches long and ½ inch or so wide, according to the size of the leg of the chicken to be caught. You can pick up a hen with this stick with the greatest of ease, especially when she is feeding.

DUCKS NEED SPECIAL QUARTERS

Does it pay to keep ducks with chickens? No, unless you have a sufficient range of the proper kind that the little ducks can be grown for about six weeks without supplying them with much bought feed, and then furnishing them for three weeks. Ducks eat four to five times as much feed as growing chickens. While they grow much faster, still if you have to supply all the feed from hatching to selling time you will find they do not pay for the trouble, when kept with chickens. There is plenty of work to do in spring and summer if you expect to succeed with chickens, so you need to avoid any more work and detail other than absolutely necessary. Ducks would mean extra work as they have to be fed more often and differently from the chickens.

I had eight ducks and two drakes last year. While I raised 100 little ducks, I sold young and old as I found that dividing my time and energy did not pay. Of course, these remarks do not apply to the keeping of a few for one's personal use, but only to the raising of a quantity for market, in conjunction with chickens. If you must try, however, let me say that growing ducks are cute. They are

GENERAL REMARKS

easily grown with few losses, are docile, not wild like chickens, and are very rapid growers, putting on half a pound a week after two weeks old. The White Pekin does not require water other than plenty to drink. It can be brought to four pounds in 8 to 10 weeks. At that weight the ducklings ought to be marketed.

Great trouble may be saved in the way of washing eggs, if you are particular to keep the nests clean, changing them as often as they require it, the result being nice clean eggs that do not need washing. Washing takes away some of the natural luster of the shell and takes considerable time if you have many to clean. So should the nests be very dirty, some stains are sure to be left.

Keep cockerels and roosters separate from hens that are supplying market eggs, as the males are of no use in the pen with these hens; in fact, rather bother them. Again, infertile eggs are better for table purposes, and will stay fresh longer. It is said that an infertile egg never rots, but simply becomes stale, which is proved by the appearance of eggs taken from the incubator after having been subjected to heat for some time. The eggs without germs have not rotted but are simply stale to the smell.

VALUE OF TRAP-NESTING

Trap-nesting hens, in order to locate good and the poor layers, is probably the only way surely to build up a strain of heavy layers, because in a large flock it is very difficult to recognize the hens that are laying and impossible to know just how often they are laying, without trap-nesting. If a flock is trap-nested, the poor layers can be weeded out, leaving a flock of all good layers from which the best can be selected for breeders, thus developing a heavy laying strain, and greatly increasing the average number of eggs laid by the flock.

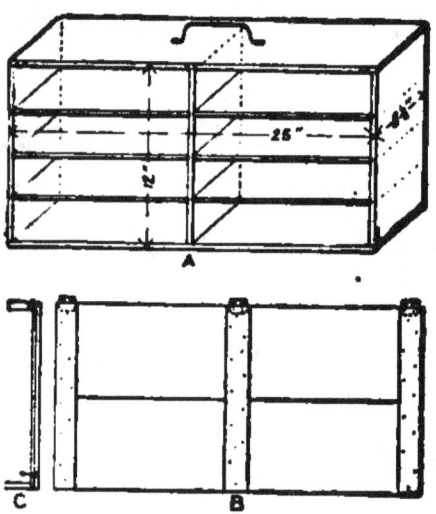

Box made of 3-16 boards for delivering cartons of eggs—capacity 16 doz.

Trap-nesting takes time, however, and until a 500 hen plant is more than a one man affair; that is, one man doing all the work, of every kind, there are some forms of work that have to be omitted, otherwise it results in nothing having the proper attention. When the time arrives that you can afford to hire help then you can start trap-nesting your flock. It will pay.

CHAPTER XXVII

DON'TS LEARNED FROM EXPERIENCE

Don't think poultry keeping is going to be easier than other ways of earning a living. It may be pleasanter and more healthful and interesting, however.

Don't get discouraged because your profits the first year or so do not reach the high mark set by some poultrymen of longer experience.

Don't expect too much the first year. You would not expect very much from any other business. Be satisfied with fair results, study, work and prepare for the next year.

Don't build any brooder or incubator house in connection with other houses. In case of fire (which, of course, manufacturers of appliances tell you is impossible with their machines), it would be bad enough to lose one house. This mistake cost me $500 once.

Don't use on an incubator or brooder any burner about which you have any doubts as to its perfect conditions.

Don't put on a cheap roof or neglect to **do** anything in the building of your houses that will help protect your birds.

Don't at any time allow your little chickens to crowd. If you use houses as planned here you will be fairly safe but if you experiment with other houses watch this point, as more deaths occur from crowding than from disease.

Don't house any chickens in a house without a tight floor, raised off the ground.

Don't forget to close up tight all houses at night to keep out marauders and the next caution will be unnecessary.

Feed Trough for growing stock.

Don't allow little chickens to get out in the early morning (when the hawks are after their breakfasts) until some person is there to protect them.

Don't move little chickens from house to house except when absolutely necessary, as the moving is likely to retard their growth. The plan of this book calls for one moving.

Don't move them to a new house without

DONT'S LEARNED FROM EXPERIENCE 167

enclosing them with wire until they get accustomed to their new quarters. I lost 200 four weeks old chicks in 1912 by moving to new houses without enclosing them. A heavy shower came up and soaked them before they could be put into their houses.

Don't let old hens, young pullets and chicks run together because some of the smaller ones will suffer from inability to get enough to eat.

Don't allow your water pans to remain empty or unclean nor your feed troughs dirty.

Don't keep ducks or geese in the same yard with chickens as they will befoul water and feed.

Don't move around your flock in such a way as to frighten them. Be gentle and they will become tame.

Don't lose your temper when the fowls fly over the fence. Build the fence higher or clip one of their wings.

Don't allow droppings to accumulate; clean twice a week.

Don't neglect to whitewash, kerosene and disinfect the houses once in a while.

Don't neglect your hens during molting time because eggs are scarce; give them all the more attention at that time and so start them laying sooner.

Don't sell any eggs about which you have any doubts as to freshness, nor any small eggs that you would not care to buy yourself.

Don't leave your houses and appliances outside subjected to the winter weather if you have any way of getting them under cover.

Notice how chicks are stepping in the trough without a cover!

Don't throw away or burn any grocery boxes as the thin boards always come in handy for feed and grit hoppers, egg crates, etc.

Don't get careless or neglectful at the two seasons of the year most conducive to carelessness—fall and winter, when hens are not

DONT'S LEARNED FROM EXPERIENCE 169

doing much—and spring, when you are busy and there are lots of growing chicks.

Don't carry over hens beyond the second year. If the flock of early hatched pullets is sufficient, pullets will pay better than hens but they must be early hatched.

Don't wait until the eggs are coming fast to find your market, or you may have to sell some eggs at a sacrifice before you locate your customers.

Don't be surprised and discouraged should your hens get to laying well during the winter and then suddenly shut off 20 to 50% due to a cold snap. Very cold weather will effect them. On January 13 my flock had worked up to an egg production of 110 increasing every day, when suddenly the temperature dropped to 14 degrees below zero, continuing zero weather for some time. My collections dropped to 44 and it was just 30 days later before the 110 mark was again reached. Hens on range with just a roosting coop are still more susceptible to the forced confinement of cold snowy weather, as they lose the effect of exercise when confined to small coops.

Don't carry over runts or poorly developed pullets in the fall. They never pay for their feed.

Don't neglect to insure your buildings and stock. Insurance gives a feeling of safety that is worth far more than the cost.

Don't fail to have a pocket flashlight lamp to look at the thermometer in your incubators during the day and for dozens of useful purposes at night.

Don't over-crowd your chickens, whether they are little chicks or old hens. They will all do better with at least comfortable space.

Don't allow anything that might interfere with the thermostat regulator to remain on top of the incubators. I burned up 240 eggs and have heard of others who smothered chickens by the heat running up in this way.

Don't under any circumstance store coops, etc., to be used the following spring to brood little chickens, where old hens can get to them and so infest them with lice. Lice are death to little chicks.

Don't let very young chicks get caught out in a shower. To prevent this

Don't allow young chicks to range until they are familiar enough with their quarters to run back in their houses. They may be made to know their house by being enclosed with wire runs for a week.

CHAPTER XXVIII

CONCLUSION

My aim has been to bring out some of the features of the business not dwelt upon very much in other books, features yet of really great importance to the beginner. I have felt that while it is very interesting to read about plants of 5,000 hens or 10,000 birds with the administration buildings, etc., it would be of more use to show the novice the details and inside workings of just such a small undertaking the average man with small capital would be able to develop. I have described only the appliances that he would need on such a small farm and have given the small details of cost and construction so it would save his time figuring out the material and cost.

I have endeavored to lay out a plan that would save his experimenting with various houses, etc.—a very expensive process. I have not tried to enter into any exposition of the chicken business from a scientific basis but have aimed to make it practical in every way. I have not claimed to make big profits, for I have not made them, nor do I believe the profits from the chicken business are any larger than those of most other lines of business which require the same attention, but I do believe that a living can be made and while making that living you can get in touch with Nature, learn some of her wonders and gain better health as well as enjoy the simple pleasures of country life.

I have gone into the marketing of eggs in detail. This though not customary is yet one of the absolutely necessary parts of the business that a beginner must know, since it is one thing to produce a commodity but another to sell it. I have given away "secrets" that have cost me many dollars to learn and which will save you just as many dollars if you can profit by my experience, and I sincerely hope that whoever reads this book may be helped, either by being kept from embarking in the poultry line if he has the idea that it is an easy way to wealth or, having decided to start, that he may be assisted through the early stages of development to a profitable living.

INDEX

A

	Page
Airing the incubator eggs	79
Adaptability	27
Air space per fowl	48

B

Bumble foot	150
Breeders need range	157
Best wire for fencing	160
Birds to select for breeders	74
Brooders and colony houses	58
Brooder and incubator house	67
Breeding and hatching	71
Backyard vs. large flock records	45

C

Clean nests	163
Crowding	166-56
Close houses at night	166
Clean-up for winter	168
Carelessness	168-93
Conclusion	171
Claims of success, mode of some	5
Cleanliness	89
Colony house with wire run	93
Cost of expressage	121
Cost per doz. of shipping eggs	121
Co-operative storage house	124
Cost of cold storage	125
Comparative incomes	5
Costly mistakes	13-15

INDEX

Capital	22
Candling	126
Cost of production of 1 doz. eggs	141
Characteristics of breeds	142
Composition of an egg	142
Common diseases of chickens	147
Comparison of small and large flocks	45-48
Caution in house building	49
Commercial hovers	60
Cost of hen hatching	78

D

Diarrhea	150
Ducks need special quarters	162
Does it pay to keep ducks?	162
Dont's learned from experience	165
Duty to your customers	113
Dimensions of egg cases	121
Dirty eggs	123
Description of laying house	51
Distribution of egg production	140-1
Details of colony house	58
Details brooder and incubator house	69
Details of hen brooder coop	65
Delivery crate	140
Distance from market	29

E

Eggs should be sold by weight	143
Effect of feed on flavor of eggs	143
Encouragements of the business	8

F

Fifty thousand eggs shipped—one claim	117
Fruit trees in chicken yard	155
Fertility of eggs	158-72-73
Furring strips	160
Fire insurance	170
Forcing pullets for laying	97

INDEX

Farmers' care of chickens	99
Fresh water—its necessity	99
Fattening broilers	101
Fattening old hens	101
Farmers' eggs—their fertility	73
Fresh eggs for hatching	73
Feeding of chickens	89
Fireless brooders	56
Feed troughs	61
Fallacial reckonings	133
Feed cost per hen	134
Flesh is sold once	104

G

Growing chicks need range	157
Green food	96
General ability	21
Good roof pays	166
General remarks	155
Growing own feed	145
Gapes	151

H

Hospital for sick chickens	151
Hen catching stick	161
Home made egg crates	119
Home made feed hoppers	96-97
Home mixed feeds	103
Hatching eggs	73
Hatching season	76
Hen hatching	85
Hens vs. incubators	77
Housing of small flocks	46
Hen brooder coop	64
Heated brooder house	54
Hotel trade	111
Hens as foragers	156
How trade grows	113

INDEX

I

Infertile eggs	163
Incubating	76
Incubators vs. hens	77
Income from my flock	19
Items of cost of 200 hen flock	40-41

K

Keep roosters from hens laying market eggs	163
Keep up your standard	112

L

Laying house details	51-54
Laying house built on the ground or raised?	49
Lice	148
Low returns to producer	110
Little chicks must be shown	90-92
Leghorns grow rapidly	101
Loss of appetite	103

M

Magnitude of the business	8
Marketing the eggs	108
Male in the breeding pen	74
Mating	75
Market	26
Molting season	34-115
Mammoth brooder stoves	70
Modern dry air cold storage	126
Methods of preserving eggs	127
Marketing the broilers	131
Marketing the old hens	135
Mites	148
Method of shipping eggs	116
Manure—source of income	155
Mash	98
Method of supplying water	106

INDEX

N

Novelty of first chickens	94
Nature's way	87-95

P

Packing eggs for shipment	120
Profits in the business	5
Patience	25
Planning to start in business	36
Private trade	112
Parcel post	116
Prepared feeds	97
Pullets lay small eggs	136
Profits—how effected	108
Price of eggs in winter	109

R

Rats	159-59
Runts	169
Routine spring work	152
Removing hovers at six weeks	95
Range necessary for hardy pullets	96
Round corners of brooders	57
Running an incubator	80
Roup	148

S

Square feet of floor space per hen	46
Separating the cockerels	63
Small colony house	66
Standard weight of fowls	144
Spray pump	149
Spring surplus	113
Storage houses	113
Small lot shipments	116
Shipping economically	117
Small eggs	123

INDEX

Storing surplus eggs	124
Scaly leg	149
Space required for 500 hens	156
Setting a wire fence	159
Scratch feed	98
Scarcity of eggs in winter	110
Small flocks compared to large	45
Supplied moisture	79
Setting leghorn hens	87
Starting in the spring	35
Small vs. large houses	43

T

Testing the thermometer	80
Teach chickens in the way they should go	92
Tests for fresh eggs	129
Thermostat	170
Trap nesting	164
The only way that pays	30

V

Very early hatches	71
Value of well known breed	33

W

Wire runs	138
Water glass	127
Weight of eggs	143
Weight of fowls	144
Wholesale price of eggs	109
Who is responsible for high prices?	111
White Pekin ducks	163
Washing eggs	163
Water	105
Willingness to work	24
Wet ground	28
What market to cater to	31
Winter eggs	99

www.ingramcontent.com/pod-product-compliance
Lightning Source LLC
Chambersburg PA
CBHW082327220526
45470CB00008B/2424